I0049288

THE
GOLDEN
EAGLE
ENTREPRENEUR

Mastering the Skies of Entrepreneurship &
Innovation with Nature's Wisdom

AMIR SALEHI | YASMIN ZIAEIAN

ISBN Paperback: 979-8-89496-420-1

TABLE OF CONTENTS

DEDICATION

To those who fly high, soaring beyond the ordinary, and daring to chase their dreams—may you always find the courage to embrace the skies and wisdom to soar in the winds of opportunity.

ACKNOWLEDGEMENT

Writing a book is never a solitary journey, and this one is no exception. We owe a debt of gratitude to many people who have supported me along the way.

First and foremost, my heartfelt thanks to our families for their unwavering support and encouragement. Their belief in our vision has been a constant source of motivation.

Thanks to our professor and instructors during our studies.

We would also like to express our deepest appreciation to founders and colleagues we have worked with and learned a lot from them.

As the author, I , Amir Salehi, would like to mention that I am deeply indebted to my wife Ghazal Jouyani for her unwavering support and motivation in writing this book.

I also want to thank my entrepreneurship professor, Håkan Boter, and my leadership coach, Fatima Nakhjavanpur. Their guidance, inspiration, and unwavering support have been invaluable to me.

As the co-author of this book, I, Yasmin Ziaeian, would like to express my sincere appreciation to Prof. Dr. Hab. Inż. Zbigniew Malara, whose invaluable mentorship and support have been instrumental in my academic journey.

Throughout my PhD studies, he has been a constant source of inspiration, encouragement, and wisdom. His leadership and expertise as a professor have not only honed personal skills but also helped me develop into a capable lecturer and researcher at the university. His impact in my life is immeasurable, and he holds a special place in my heart.

We say a big thanks to all the entrepreneurs, venture capitalists, and corporate leaders who shared their stories and expertise with us! Your real-life experiences and advice have made this book way more relatable and helpful. We're super grateful for your willingness to share your wins, losses, and lessons learned - it's been a total game-changer for us!

ABOUT THIS BOOK

"**W**ant to soar in the world of entrepreneurship? This book 'The Golden Eagle Entrepreneur' is for you! Whether you're a startup founder, a venture capitalist, corporate leader, or just a player in the startup scene, we've got the goods to help you navigate the wild ride of entrepreneurship. Get ready for practical tips, real-life insights, and strategies to help you fly high as a Golden Eagle Entrepreneur!"

Inspired by the majestic golden eagle, the book explores how the bird's behaviors and strategies can be translated into effective entrepreneurial practices.

The golden eagle symbolizes ambition, power, vision, strategic decision-making, and resilience. By understanding and emulating these qualities, entrepreneurs can enhance their ability to recognize opportunities, innovate, and succeed in their ventures.

Meet the author, Amir Salehi, a seasoned entrepreneur, venture capitalist, and startup mentor with over 15 years of experience in entrepreneurship and innovation! He has a unique approach, inspired by his love for nature and the golden eagle, that'll give you a fresh spin on how to crush it as an entrepreneur. With his guidance, you'll learn how to spread your wings and soar to new heights!

The co-author, Yasmin Ziaeian, is the other half of this dynamic duo! She's an academic with a solid background in management and strategic studies, which perfectly complements Amir's real-world experience and insights. Together, they've created a book that's like the ultimate entrepreneurship cheat sheet - it's got the theory to back it up and the practical tips to make it happen! Yasmin keeps things grounded and real, so you get a balanced view of what it takes to succeed in the entrepreneurial world.

Who Should Read This Book?

- **Entrepreneurs** seeking innovative strategies to grow their businesses.
- **Venture Capitalists (VCs)** looking for insights into entrepreneurial practices and success stories.
- **Corporate Leaders** aiming to foster an innovation mindset within their organizations.
- **Business Professionals** who desire to enhance their entrepreneurial strategic decision-making and leadership skills.
- **Players in the Startup Ecosystem**, including incubators, mentors, and accelerators, who support and nurture emerging businesses.

What You Will Find Inside

This book is the ultimate guide to entrepreneurship - it contains a mix of theory, real-life stories, and personal experiences from the authors. They're covering all the critical aspects, from spotting opportunities and innovating, to taking risks and bouncing back from failures. Each chapter dives into a different part of the entrepreneurial journey, so you get a complete picture of what it takes to succeed. It's like having a mentor in a book!

The Golden Eagle Entrepreneur serves as a source of inspiration and a practical tool for those looking to navigate the entrepreneurial landscape with the wisdom and innovation exemplified by the golden eagle. It is ideal for entrepreneurs, VCs, corporate leaders, incubators, mentors, and other key players in the startup ecosystem seeking to soar to new heights of success.

CHAPTER ONE

INTRODUCTION

"Fly with the wings of the eagle, and let
the sky be your guide." - **Rumi**

Entrepreneurship is more than just the mere act of starting businesses;
it's about identifying opportunities, taking calculated risks, and crafting
innovative solutions to meet market demands.

This book explains the core of entrepreneurship by drawing insightful
parallels with the majestic golden eagle. With its exceptional vision,
strategic prowess, and unwavering determination, the golden eagle
provides profound lessons on the behaviors and mindsets that drive
entrepreneurial success.

On the Essence of Entrepreneurship

Entrepreneurship is on the rise, and institutions worldwide are stepping up to support and encourage it. Universities, business incubators, accelerators, startup studios, venture capital companies, and angel investors are all becoming key players in helping future entrepreneurs learn the ropes. They're like a team of mentors, providing the tools and know-how to help newbies turn their ideas into reality.

Entrepreneurship is a complex process often fueled by a vision for change, a passion for certain products, services, or customers, or the desire for independence.

Entrepreneurs first discover opportunities before exploiting them; this entire cycle of opportunity-driven activity is determined by different factors, both within and outside the entrepreneur. Knowing why people take an entrepreneurial path goes a long way in creating a successful business environment. Therefore, entrepreneurship is very important. It drives economic growth, creates jobs, and promotes innovation, all of which help make the world a better place.

Entrepreneurship is all about turning big ideas into real-life game-changers that can actually shift society. When entrepreneurs innovate, they shake up the market, creating healthy competition that drives productivity and keeps things fresh.

Characteristics of Successful Entrepreneurs

So, what makes successful entrepreneurs ? Well, they've all got some key traits that help them navigate the wild ride of starting and running a business.

Of course, no two entrepreneurs are exactly alike, but there are some common qualities that many of them share. Think of it like a recipe for success - they've all got a dash of this and a pinch of that, which helps them overcome obstacles and come out on top.

- **Passion:** Intense drive and motivation to surmount obstacles and persist in their endeavors.
- **Vision:** Ability to think strategically, identify emerging trends, predict market changes, and conceptualize creative solutions to meet unfulfilled demands.
- **Adaptability:** Flexibility to adjust strategies, embrace new technologies, and pivot business models when necessary in a rapidly changing business environment.
- **Risk-Taking:** Comfortable with taking bold risks and making decisions despite potential challenges and setbacks.
- **Resilience:** Capability to learn from setbacks, bounce back from disappointments, and maintain a positive mindset.
- **Self-Assurance:** Confidence in their skills and choices, empowering them to be proactive, take action, and convince others to back their endeavors.
- **Diligence:** Willingness to put in long hours and make sacrifices, driven by a strong sense of personal responsibility.

- **Continuous Learning:** Commitment to staying updated on industry trends, seeking out new skills, and remaining open to feedback and mentorship.
- **Networking:** Understanding the value of building strong networks and relationships, actively seeking opportunities to connect with industry experts, potential customers, and strategic partners.
- **Emotional Intelligence:** Ability to communicate effectively, build rapport, inspire, and motivate their teams.

Types of Entrepreneurship

Entrepreneurship encompasses various types, each with its own unique focus and objectives:

Small Business Entrepreneurship

- This involves starting and operating a business on a smaller scale.
- It focuses on serving local or niche markets and providing goods or services to meet specific customer needs.
- It aims for stability, profitability, and long-term sustainability rather than rapid growth or disruptive innovation.

Scalable Startup Entrepreneurship

- This refers to ventures with the potential for rapid growth and scalability.

o Entrepreneurs strive to develop groundbreaking products or services that can be readily adopted on a large scale.

o It often seeks external funding, builds high-growth teams, and focuses on capturing significant market share. Examples include technology startups and disruptive businesses in industries like e-commerce or software development.

Social Entrepreneurship

o This involves creating businesses with a primary focus on addressing social or environmental challenges.

o It aims to generate positive social impact alongside financial sustainability.

o It develops innovative solutions to societal problems such as poverty, education, healthcare, or environmental sustainability.

Corporate Entrepreneurship

o This involves entrepreneurial activities within established organizations.

o Corporate entrepreneurs identify and pursue new opportunities, develop innovative products or services, and drive change and growth within their companies.

o It navigates organizational structures, manages resources, and balances innovation with existing operations to foster a culture of entrepreneurial thinking within the company.

Lifestyle Entrepreneurship

- o This involves starting and operating a business that aligns with an individual's personal interests, values, and desired lifestyle.
- o It prioritizes personal fulfillment and work-life balance over rapid growth or high profit margins.
- o Its ventures often revolve around hobbies, passions, or lifestyle choices, providing the entrepreneur with independence and flexibility.

Entrepreneurial Process

So, do you want to turn your big idea into a reality? Well, here's the lowdown - entrepreneurship is a step-by-step process that takes your idea from zero to hero.

There are two main versions of this process: the structured Business School Process and the dynamic Intuitive Process.

Business School Process of Entrepreneurship

1. **Opportunity Recognition:** Identify potential business opportunities based on market needs and customer demands.
2. **Feasibility Analysis:** Assess the viability and potential of identified opportunities.
3. **Business Planning:** Create a comprehensive business plan outlining goals, tactics, marketing methods, operations, and financial forecasts.

4. **Resource Acquisition:** Obtain necessary resources, including funding, skilled staff, and physical assets.

5. **Implementation:** Execute strategies by setting up infrastructure, establishing procedures, and launching products/services.

6. **Marketing and Sales:** Implement marketing and sales strategies to promote offerings, attract customers, and drive sales.

7. **Growth and Scaling:** Expand into new markets, diversify offerings, form strategic partnerships, or seek external funding.

8. **Continuous Improvement:** Evaluate performance, gather customer feedback, and iterate strategies to adapt to changing market conditions.

Intuitive Process of Entrepreneurship

1. **Opportunity Recognition:** Identify opportunities based on personal experiences or sudden insights.

2. **Rapid Prototyping:** Develop a rough version of a solution (prototype or Minimum Viable Product - MVP).

3. **Market Validation:** Gather informal feedback from potential customers and test the MVP in the market.

4. **Lean Development:** Emphasize essential features, use agile methodologies to release updates frequently.

5. **Customer Feedback Loop:** Engage with early adopters to collect ongoing feedback, implementing changes swiftly.

6. **Market Testing:** Conduct small-scale launches or beta tests to assess market response and refine the product.

7. **Scaling:** Gradually scale production, distribution, and marketing efforts based on proven demand and feedback.

8. **Resource Mobilization:** Seek additional funding, talent, or technology as needed to support growth.

9. **Pivot and Adapt:** Regularly review market trends and customer feedback, pivoting business models or features as necessary.

Entrepreneurship and Strategy

Starting a new business from scratch? That's like building a house from the ground up! And if you're a technopreneur, you need to be a double threat - a master of tech and a whiz at managerial skills.

Frameworks like Sara Sarasvathy's effectuation theory, Steve Blank's customer development, and Eric Ries's lean startup promote a fail-fast-and-fail-cheap mindset, while also emphasizing strategic thinking to identify and seize opportunities.

For technopreneurs who are more tech-savvy than business-savvy, this strategic approach is crucial. Luckily, the lines between entrepreneurship and strategic management are blurring, making it easier for founders to learn and adapt.

Think of entrepreneurship as a powerful tool that can drive innovation and growth within existing organizations.

Concepts like corporate entrepreneurship and strategic entrepreneurship have been coined to underline this mixture, emphasizing the role start-ups can play in continued innovation and success of large organizations.

Large companies are looking to form alliances with more start-ups these days, as they realize the fact that strategic entrepreneurship can assist them in realizing their bottom-line objectives as well as their broader strategic aspirations.

Today, the business ecosystem includes support systems that help entrepreneurs navigate the challenges of starting a business, accelerators, mentorship programs, and networking platforms.

Strategy and innovation together drive economic and job growth through entrepreneurship, setting the stage for a more innovative and prosperous future.

Entrepreneurship is like a shot of adrenaline for organizations! It brings in fresh ideas, innovative thinking, and new opportunities that can shake things up and drive growth. When entrepreneurs are involved, they can inspire new strategies and initiatives that might not have been considered otherwise.

Strategic management gives a guideline on how to make choices and set resources in the direction of some stated long-term goals, thus helping entrepreneurs direct resource allotment, goal setting, and growth planning effectively.

In addition, Entrepreneurial Orientation—being a firm's tendency to be innovative, risk taking, and proactive opportunity seeking—is very critical both in entrepreneurship and strategic management.

By its name, strategic entrepreneurship combines the best from the both worlds: focusing on entrepreneurial actions that include spotting and capturing opportunities in the dynamic, competitive environment while blending with strategic decision-making and aligning it with the future corporate strategy.

> **"Look deep into nature, and then you will understand everything better." — Albert Einstein**

Inspired by Nature

Nature is like the ultimate mom - who feeds and teaches us valuable lessons through the amazing diversity of ecosystems.

And let's be real, wildlife has been a major inspiration for human innovation. By studying the natural world, we've been able to develop an advanced society and technology. For example, submarines are inspired by marine wildlife, while missiles and fighter jets take inspiration from falcons and hawks, respectively.

Wildlife has some serious management skills to teach us. Take bee colonies and wolf packs, for example. Their hierarchical leadership structures show us how to make strategic decisions and lead effectively.

Cooperative behavior within animal groups provides models for teamwork and cooperation. These further ways of enhancing internal communication and resilience in business environments can be motivated by the diverse adaptability to habitats and the animal communication systems.

The golden eagle - a bird of prey that's not only majestic but also a master strategist! By studying its hunting and survival tactics, entrepreneurs can gain a fresh perspective on how to refine their own strategic approaches. This bird's impressive skills can reveal new ways to enhance strategic thinking and decision-making processes.

From scanning the landscape to swooping in on opportunities, the golden eagle's tactics can inspire innovative solutions for business leaders. Who knew that observing a bird's behavior could lead to breakthroughs in strategic planning?

CHAPTER TWO

THE GOLDEN EAGLE ENTREPRENEUR

"You were born with wings, why would you prefer to crawl through life?" — **Rumi**

To truly understand what it means to be golden eagle entrepreneurs, we first need to explore the world of the golden eagle itself.

This majestic bird, with its keen vision, strategic hunting techniques, and relentless drive, offers profound insights into the behaviors that define successful entrepreneurship. In this chapter, we will get into the brain of the golden eagle, and translate its qualities into the context of entrepreneurship.

On This Stunning Bird of Prey

Common Name: Golden Eagle

Scientific Name: Aquila Chrysaetos

Average Life Span: 30 Years in Nature

Length: 27 to 38 inches

Wingspan: 72 to 87 inches

Weight: 6 to 15 pounds

The golden eagle, known as the ultimate bird of prey! With its striking appearance and impressive wingspan, this bird is a force to be reckoned with. But it's not just its looks that make it remarkable - its unique behaviors and hunting techniques are equally fascinating!

With a wingspan of up to 87 inches and a length that exceeds 30 inches, the golden eagle is among the largest birds of prey.

The bird's feathers are mostly dark brown with some upper parts of yellowish brown. The tail is lighter but has darker tips. The striking part of the bird, however, is where the name originates: the top of the head and the neck of the golden eagle are golden-brown, glowing like a crown of a king!

These hackles shine in light, sometimes appearing white when the bird is flying in the sunshine. The eyes are hazel-brown, and the beak is gray. The powerful legs are feathered with yellowish brown feathers extending to pale yellow talons, equipped with sharp black claws, enough to snatch any kind of prey.

An interesting fact about the golden eagle is the size difference between males and females. The female golden eagle is about 30% heavier and has wings that are 10-12 percent larger than the male.

Juveniles can be distinguished by the white patches of feathers under and on top of the wings, as well as the base of the tail. These feathers darken as the golden eagle ages and matures into adulthood.

Properties

The golden eagle is equipped with advanced tools and weapons for hunting, competition, and survival.

First and foremost, its eyes are the primary tool for spotting prey. The golden eagle's eyes are relatively large, taking up 50% of the head, and can spot objects from 4-5 times farther away than human eyes. Positioned on the sides of the head, the eyes provide a 340-degree field of vision.

Moreover, the golden eagle has superior color recognition, perceiving five colors compared to three for humans. This extraordinary eyesight helps the golden eagle spot and recognize objects from a distance, even when camouflaged.

Due to the author's frequent presence in nature, he has been privileged to watch this amazing bird near the mountains where it belongs:

"Recognizing a flying bird from a distance with the naked eye can be challenging, but we can identify a soaring golden eagle by its unique dihedral angle and finger-like feather tips.

The wings enable efficient and fast flight. On one occasion, while I was driving, I saw a juvenile golden eagle soaring at a low altitude.

As I pulled over to capture a photo, the eagle, feeling threatened, soared higher in less than a minute. This demonstrates the golden eagle's efficiency in gaining altitude with minimal effort and just without fapping wings."

The broad wings also help handle larger prey without expending much energy. The tail complements the wings, aiding in acceleration and acting as a speed brake.

The size of the bird and its wings make it one of the fastest birds on earth, rivaling the peregrine falcon. Both birds use their speed to attack prey from above, diving with tucked wings.

The legs and talons of the golden eagle are used to snatch prey and defend the nest and territory. The legs provide grip strength, and the talons are sharp enough to capture and kill prey.

They also help the bird land and are usually tucked close to the body during flight. The hooked beak helps clean and eat prey. These physical attributes allow the golden eagle to build nests in unreachable places, mostly on higher cliffs. The golden eagle also possesses sophisticated intelligence, including mental ability and high decision-making power.

The Life Cycle

Courtship is one of the most spectacular scenes in nature, for male and females to take advantage of windy days by soaring into the sky in what Watson called "sky dancing".

They climb to higher altitude with the male usually getting higher and making several stoops to meet the female.

The female shows its talons and they tangle and grip, rolling together. They mate on the ground and the result is usually 2-3 eggs, though rarely four.

Incubation is for up to 45 days, with over 80% of it done by the female. After hatching, the nestlings are fed mainly by the female until they can tear up food alone.

The chicks stay in the nest for 70-80 days before fledging. They exercise their wing muscles and finally make a descent flight from the nest.

Parents feed them and teach them to hunt until they become independent hunters. It takes about three months after fledging for the golden eagle to attain the same quality flight as the adults and become independent.

Four years are required for it to reach sexual productivity from the juvenile stage. Some juveniles may starve in the first two years due to poorly developed hunting skills. But once mature and capable of efficient hunting, they often live close to 30 years.

Diet and Home Range

So, what's on the menu for a golden eagle? Well They feed on small mammals like squirrels and hares, but they're not afraid to go after bigger prey like deer, mountain goats, and even other predators like wolves and foxes. They're like the ultimate hunters, taking down whatever they can catch!

Preying on reptiles like snakes and lizards has been reported, and it is also known to drop turtles from the sky to break their shells. In some regions, they feed on spent fish from the rivers.

They also hunt other birds including ducks, partridges, grouse and sometimes geese.

As Jeff Watson reports, a golden eagle's territory normally covers from 10 to 70 square miles (25 to 180 sq. km) and most often there are several nest sites, but some of them can be favored more than others.

Having dozens of nest sites, only one or two can be used commonly. Multiple series in one territory are a way of dealing with contingencies.

Nests are usually built on mountains or in the oldest trees in the area. Cliff nest sites give them an advantage since they can ride the wind and get all the upward drafts to make it easy for the eagle to soar into the nest with prey.

Behavior and Hunting Techniques

So, when a golden eagle goes on the hunt, it's like a stealthy ninja! It's all about sneaking up on its prey, grabbing it fast, and then consuming quickly and discreetly.

Golden eagles adjust their hunting style to fit the prey and terrain. They're adaptable predators, switching up their tactics to catch their next meal.

Of course, the first process is spotting the prey, possible due to its keen eyesight and silent flight achieved when soaring at extreme altitudes.

From such heights, the golden eagle focuses intensely on its target after finding it from above in several circular rounds.

A successful hunter, the golden eagle often meets some possible prey. Nevertheless, it is similar to most wildlife hunters that consider the gain in energy by the prey higher than the loss in energy used during the hunt. The pre-hunt calculation is vital for its success.

That's why the golden eagle does not prey on everything that it finds. Golden eagles will either soar high and dive or fly low to the ground depending on circumstances.

Low flights, for instance, work best when hunting grouse or prairie dogs, while hares are chased and birds in flight facing a vertical high-speed stoop attack in mid-air.

Documentaries like the 1970s "El Hombre y la Tierra," of Spanish TV, showed how golden eagles used a drag-and-drop technique, letting mountain goats drop from high altitudes while hunting them. The process is risky and somewhat dangerous, with the result of a number of such attempts resulting in the injury or even death of golden eagles from goring deer upon attack.

Love & Watson (1990) found that in many instances, golden eagles in the UK have been killed by the antlers of deer that they attacked.

They discovered that, particularly with juvenile birds, due to lack of experience in hunting, many lose their lives during an attack.

In one such instance, a camera trap set up by biologists to capture pictures of tigers in Russia captured a picture of a golden eagle attacking a Sika deer.

Juveniles practice first with simulated attacks from their parents. The training of the juveniles begins when parents drop dead prey for the juvenile to catch. In most cases, the first trials do not succeed, but through a bit of practice, juveniles learn how to capture and kill their prey.

The golden eagle learns how to fly with the wind, how to find prey, and how to use various hunting methods in a heuristic manner.

More experienced birds take bigger risks by preying on larger species. In this case, creativity steps in where most conventional hunting techniques fail. Juveniles may commit calculation errors, but they learn from these mistakes by perfecting their techniques.

The Golden Eagle Strategy

Exploration into the life of golden eagles demonstrates that the bird copes with a very high level of contingency in which it first locates its prey (exploration) and then focuses and snatches the prey (exploitation).

The golden eagle hunts start by exploring opportunities in its environment and this is done by scanning the ground and scrutinizing everything from above.

Unusual tools to explore existing opportunities are the sharp eyes. The Golden Eagle circles repeatedly, scanning the area in a 360-degree move during the search.

This kind of extensive collection and processing is what we term as the true, valid, and rational intelligence shown by the golden eagle in its decision making.

The golden eagle has the ability to collect, process, and analyze information at a very fast rate, this includes opportunity assessment. The golden eagle also plans its move, deciding when to make a move.

Its mindset further complements hunting efforts: instinctive risk-taking, high tolerance for ambiguity and contingency, flexibility, innovativeness, and efficiency. For example, mountain goat is not a regular food for golden eagles as it is much larger in size, but the courage to take a risk along with innovativeness of a drag-and-drop technique in hunting large animals like this make the golden eagle initiate the hunt with decisiveness.

Efficiency matters or no energy will be left. The switch from exploration to attack is fast. The golden eagle does not over analyze and takes an efficient and bold action. It is crucial to get the most by spending the least energy possible by leveraging the wind and momentum.

Prey focus is a vital part of the hunt; after separating a prey, the golden eagle locks on the prey and focuses on the techniques of hunting the targeted prey.

This allows the golden eagle to manage contingencies better since it knows the escape strategies of every single prey in the mind from past experiences.

After the attack, it must also be flexible enough to orient direction if the prey makes evasive maneuvers.

The hunt is highly ambiguous and contingent; not every attempt pays off. But after each failure the golden eagle gains a new experience that is helpful in the next hunt. Failure in hunting is a part of life of the golden eagle.

The golden eagle has multiple nest sites, which it uses frequently. Although some nests are more favored, the golden eagle uses a variety of nest sites and diversifies the territory.

The nests are often built on high cliffs, making them inaccessible to predators and giving the eagle a strategic advantage. This demonstrates the golden eagle's future foresight. Besides occasional migration, multiple nests indicate that the golden eagle builds its future, based on experiences and instincts. The vast home range is critical to the survival of the golden eagle. The network of aeries within this large territory allows the golden eagle to maintain its dominance and ensure a steady supply of resources.

The Golden Eagle Entrepreneur

Just because a bird has wings doesn't mean it's a golden eagle, right? Same thing with entrepreneurs - just because someone starts a business doesn't mean they're golden eagle entrepreneurs.

Golden eagles are distinct in their prowess, intelligence, and strategic approach to survival and success.

Similarly, golden eagle entrepreneurs stand out for their unique blend of vision, resilience, and innovation.

These entrepreneurs do not just start businesses; they navigate complex landscapes, take risks, and turn challenges into opportunities with a strategic finesse akin to the majestic bird.

To understand what sets golden eagle entrepreneurs apart, we need to explore the golden eagle's brain and examine their behaviors and strategies.

Let's investigate the specific traits and actions that define a golden eagle entrepreneur.

The Golden Eagle Intelligence

The golden eagle's intelligence is the brains behind its operations, driving its every move and decision, especially when it comes to its hunting journey.

Just like the golden eagle's sharp mind to hunt and thrive, entrepreneurs have a mindset that helps their business soar to new heights and stay ahead of the game!

Golden eagle intelligence is the driver of entrepreneurship and makes the real distinction between people who possess that type of intelligence and those who don't.

This intelligence is actively working during the whole process of entrepreneurship from exploration to exploitation of the opportunity.

When mapping the golden eagle behavior and comparing it with the successful entrepreneurs we come up with the following components of the golden eagle intelligence. Entrepreneurs with golden eagle intelligence possess:

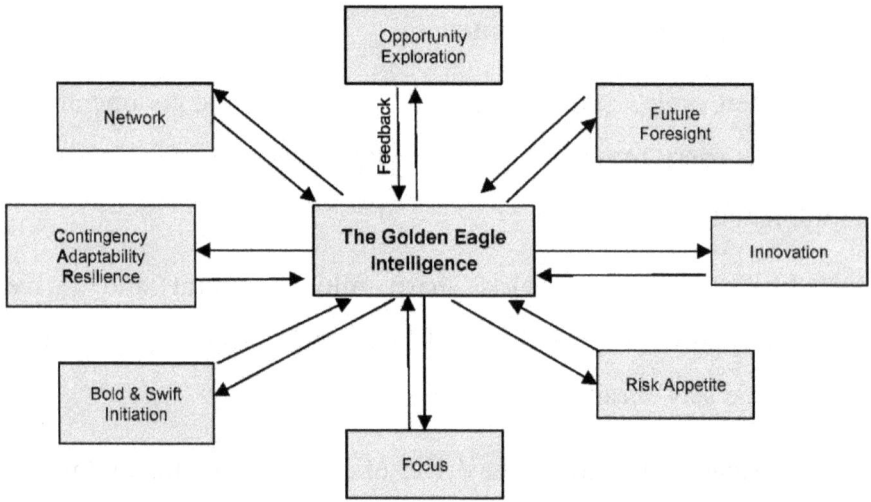

Opportunity Exploration

Golden eagles are always on the hunt for their next opportunity - they're like the ultimate opportunists, constantly scanning for their next meal!

This trait is fundamental to its hunting strategies. Golden eagle entrepreneurs are similarly opportunity-oriented. They define the correct opportunity, recognize it, analyze it, and eventually exploit it. They are always on the lookout for the right moment to strike.

Future Foresight

Golden eagles have a huge territory and are always on the move, which helps them stay one step ahead of future challenges.

They consider prey scarcity and future weather conditions, creating multiple nest sites and migrating when necessary.

Similarly, entrepreneurs need to be future-focused, anticipating and creating new opportunities for their startups to thrive. It's all about building a sustainable future, not just living in the moment!

Innovation

The golden eagle's innovation is evident in its hunting strategies, such as using the drag-and-drop method to hunt larger prey like mountain goats.

Innovation is the heart of entrepreneurship, distinguishing entrepreneurs from other business founders.

Innovative entrepreneurs create new opportunities and exploit them, viewing innovation as an integral part of their mindset.

Risk Appetite

Golden eagles are fearless hunters, taking significant risks to snag bigger prey like deer, even if it means getting hurt by their sharp antlers!

Despite these dangers, golden eagles manage risks with innovation. Similarly, entrepreneurs must take risks to introduce new innovations. Risk-taking and innovation are essential for entrepreneurial success, as without risk, there can be no innovation.

Focus

Once a golden eagle recognizes an opportunity, it locks onto a single prey and decides to capture it!

This focus is critical, marking the transition from exploration to exploitation. Successful entrepreneurs also demonstrate a high level of focus, concentrating on one task at a time, just as the golden eagle focuses on one prey at a time.

Bold Initiation

With speed, sharp talons, keen eyesight, and intelligence, the golden eagle swiftly seizes its prey, acting decisively before the opportunity slips away.

This bold initiation is essential, despite occasional failures. Each attempt, successful or not, refines the eagle's approach.

"Entrepreneurship is neither a science nor an art. It is a practice." - Peter Drucker

In entrepreneurship, talk is cheap - it's all about taking bold action! Spotting an opportunity is just the start, the critical step is to act boldly and decisively. Go for it! Entrepreneurs must embrace action, learn from mistakes, and keep refining your strategies until you nail it!

Network

The golden eagle's vast home range includes several nest sites that are used according to the conditions.

This network helps cover more ground while hunting, enhancing survival and hunting success.

Entrepreneurs also need to build strong networks of individuals and corporate partners. A well-established network aids in better performance from exploration to exploitation of opportunities.

Contingency, Adaptability, & Resilience

The golden eagle's life demands careful contingency planning due to the instability and unpredictability of nature. It must adapt to prey behavior, changing weather, and other unforeseen challenges.

Similarly, entrepreneurs face contingencies and uncertainties. Effective contingency planning and adaptability are vital for startup survival. Entrepreneurs must be resilient, making the right decisions when challenges arise to navigate the unpredictable business landscape.

Failure and the Feedback Loop

Whether it's a golden eagle missing its prey or an entrepreneur's failed launch, mistakes are actually a good thing! They give you valuable feedback to learn from and come back stronger, wiser, and more likely to succeed next time!

Each hunting attempt or business action, whether successful or not, provides insights that refine future strategies. This iterative process of learning from each experience ensures that both the eagle and the entrepreneur become more effective and adaptable over time.

CHAPTER THREE

OPPORTUNITY EXPLORATION

"Whether one moves slowly or with speed, the one who is a seeker will be a finder." - Rumi

Entrepreneurship starts with exploring opportunities. It's important for golden eagle entrepreneurs to recognize and act on opportunities. Just like the majestic golden eagle that soars high, scanning the landscape with sharp eyes, these entrepreneurs are born to find and grab potential business opportunities.

They're great at noticing problems and gaps, future trends, and unmet needs, turning them into innovative ventures. In this chapter, let's get into the nitty-gritty of how golden eagles and entrepreneurs sniff out new opportunities!

What is Opportunity?

Dictionary defines the term *opportunity* as "*a favorable juncture of circumstances*" and "*a good chance for advancement or progress*".

There are also other definitions, which all take opportunity as a chance that might lead to positive outcomes.

In sports, players seek opportunities to score or beat the opponents, people within an organization work hard and try to take advantage of opportunities to get promoted, in the stock market a possible good investment is an opportunity, and in many other aspects of our lives opportunity is the favorable situation we are looking for.

Tying *chance* to opportunity as per some definitions, might misguide us in understanding the real nature of opportunity.

The condition is not an opportunity, if we just wait for the promising results through luck. Therefore, the main difference between these two terms lies behind *potential action* and *control.*

Opportunities are useless if you don't take action! It's all about making a move and being in control of your decisions - that's what separates luck from a real opportunity.

You usually have control over opportunity; you can recognize it, create it, develop it, miss it, or exploit it. Whatever it is, you have control over a decision in opportunity.

The soccer player can score or miss the opportunity based on his/her talent. A good fly fisher matches the hatch and uses the correct fly and creates the opportunity by making a good cast, good choice, good fly presentation, and a good landing of a fish.

A rolling dice, on the other hand, is not under your control once you rolled it and any result you get from it, would be a chance.

But still if you are playing a backgammon or any other game using dice, to some extent you have control over the game, as the choice of your strategy matters. This is where you deal with opportunity rather than luck.

In the book "Extreme Ownership", Jocko Willink & Leif Babin presented the leadership lessons from the US army and while mentioning Chris Kyle, the US sniper in Iraq, they indicated that Kyle was lucky because he usually made his own luck with his own talent.

While a soldier or a sniper might be lucky sometimes, creating luck with his actions or use of talents belongs to opportunity recognition or development, not chance or being lucky.

Entrepreneurial Opportunity

Opportunity is the heartbeat of entrepreneurship! It has been the central issue in hundreds of studies by leading scholars.

Almost all researchers on the subject of entrepreneurship agree that opportunity is the basis of entrepreneurial activities.

In order to define entrepreneurial opportunity, we would like to first borrow Robert Baron's definition, which is *"discerned and unexploited means (or as Shane puts it, means-ends relationship) by which entrepreneurs create economic value."*

In this definition, recognition of new means is one of the main elements that is required for defining entrepreneurial opportunity and hence *newness* plays an important role in entrepreneurial opportunities.

The newness as a factor is the base of innovation which is vital for entrepreneurship. Newness of the means or means-ends relationship makes the real distinction between entrepreneurial opportunities and other forms of opportunities people might perceive.

The exchange rate difference in different markets is an attractive situation to make money through arbitrage and it is an opportunity.

Meeting a key business person in a business show and making a new deal is definitely an opportunity. Finding a key engineer for your company can also be an opportunity, but none of these examples belong to entrepreneurial opportunity and its definition as they all explore and exploit an existing phenomenon and create value through creation of optimization of existing platforms.

However, making a new cloud based platform, which is powered by Artificial Intelligence and selling it as a service to companies is an entrepreneurial opportunity.

It all starts with opportunity recognition, which is an iterative process in entrepreneurship. Searching may be either passive (entrepreneurial alertness) or active, and events may be recognized as opportunities or not.

The speed and accuracy at which an opportunity is recognized are influenced by external factors and the entrepreneurial bias.

However, recognizing an opportunity is just a first step and very far from a final decision. Scott Shane and his co-authors found in their research that entrepreneurship is all about exploring and exploiting opportunities while regarding the human factor as a vital element in this process.

Opportunity Exploration

The exploration of entrepreneurial opportunities involves a process comprising opportunity recognition, opportunity evaluation, and opportunity development. The very first step in entrepreneurship occurs when an individual recognizes an opportunity and seeks to realize future rewards through entrepreneurial action.

Entrepreneurial opportunities are not perceivable to everyone; people have different tendencies and perspectives toward various phenomena.

For example, while many people might be fascinated by the advent of VR headsets, a particular individual might see the potential to create a VR-based game designed to aid in rehabilitation through physiotherapy.

This difference in perception is one of the foundations of entrepreneurship. As Schumpeter noted, if everyone had the same entrepreneurial insight, there would be no entrepreneurial value because everyone would pursue the same opportunities, dividing the profit to the point where entrepreneurial gain and value would be eliminated.

Therefore, the human factor plays the key role in exploring opportunities. It is individuals who recognize, evaluate, and make the final decision on pursuing opportunities through the act of entrepreneurship.

This personal touch ensures that entrepreneurship remains dynamic and innovative, driven by unique insights and the diverse experiences of individuals.

Entrepreneurial Opportunity Recognition

Opportunity recognition is a cognitive process that helps potential entrepreneurs determine whether certain phenomena can create wealth and value.

As a human-centered process, individuals have different approaches to recognizing opportunities.

While opportunities are objective, their recognition is subjective, relying on both human and external factors. Opportunity recognition occurs in two primary forms: active search and entrepreneurial alertness.

Active Search

Active search for entrepreneurial opportunities is akin to actively hunting for something specific, like looking for food in a fridge or searching for a raptor with a camera in hand.

For a golden eagle, it means soaring high in the sky, actively scanning for prey. Similarly, entrepreneurs engage in active search by collecting information, seeking business partners, and attending events.

As the co-founder and CEO of an AI accelerator, the author expressed:

"We invested in AI startups and ran multiple programs. I always made time for office hours to allow potential clients and customers to talk to me directly. Our customers were potential founders. During these sessions, I met many individuals who were actively searching for entrepreneurial opportunities, looking for co-founders, ideas, or knowledge on how to start their own business. Although not everyone who is actively searching becomes an entrepreneur, this is one of the very first stages in entrepreneurship."

Active search is an ancient human trait that started from the age of neanderthals. In the business world, part of our daily life is also spent on active search and is not limited only to those who want to start a business.

Experienced entrepreneurs and leaders of larger firms are also actively looking for entrepreneurial opportunities, often more actively than individuals.

As an example, VCs and accelerators can create better opportunities, when they set up a scout team with a plan to actively look for innovators instead of waiting for founders to submit an application on their website. This proactive approach to seeking out opportunities is essential for staying competitive and innovative in the business world.

Entrepreneurial Alertness as a Passive Search

Entrepreneurial alertness requires having a curious mind that's always on the lookout for certain events.

This is a form of passive search where you are not actively looking but remain open to opportunities.

In the book *Thinking, Fast and Slow*, Daniel Kahneman divides our thinking into two systems: System 1, which is fast, powerful, impulsive, and runs automatically; and System 2, which is rational, slow, calculative, and cautious.

Entrepreneurial alertness is based on System 1, which operates like a radar, perceiving familiar phenomena and recognizing them as opportunities.

To be alert to something, you should have experience or familiarity with that situation so that System 1 can spot the event and recognize it as an opportunity.

Familiarity with a topic increases sensitivity and triggers a System 1 response. For example, when you are in business, you become sensitive to the logos of your own company, competitors, or suppliers on a billboard.

When familiar with a topic, you respond to specific keywords in the news or ads, making you alert to potential opportunities even in unrelated situations.

Here is author's experience that illustrates this sensitivity and alertness:

"When I see the word 'fly,' I am immediately drawn to it, waiting to see 'fishing' after it. It happened to be an airline ad, but my eyes go directly to the word 'fly' first because I am a fly fisher and passionate about it. When I had my startup, Lar Fishing in Sweden, I was often attracted to the word 'Lars,' a common Swedish name. Pausing on the word corrected my mistake, a process governed by System 2. This kind of alertness can apply to any topic or event. Entrepreneurial alertness is also our sensitivity to events, particularly those relevant to entrepreneurial opportunities."

Prior business ownership develops a sense of opportunism in individuals. Whether the business was a success or failure, active or closed, and regardless of its size, it fosters a sense of alertness.

Entrepreneurs become sensitive to certain words or topics related to their business experiences, such as contracts, funding, hiring, and product development.

Entrepreneurial alertness is based on the entrepreneur's bias, which can lead to opportunity recognition.

This alertness remains active throughout the entrepreneur's life. Whether running a business, on vacation, or jogging in the morning, once you start your first business, you become sensitive to related events for the rest of your life.

Recognition Stimuli

External Factors

External factors also play a crucial role in entrepreneurial opportunity recognition. The environment in which individuals work and live significantly impacts their ability to identify opportunities.

An event might be recognized as an opportunity for an individual in the U.S, while the same person might not see any opportunity in that same event elsewhere. Factors such as ease of doing business, regulations, culture, industry trends, and technological advancements influence this process.

In strategic management, tools like PESTLE analysis are used to evaluate these external factors.

PESTLE analysis examines the Political, Economic, Sociocultural, Technological, Legal, and Environmental aspects to understand the impact of environmental risks and circumstances.

When evaluating an event, an entrepreneur's mind considers these factors, weighing them carefully to make informed decisions. This comprehensive evaluation helps in identifying and exploiting opportunities effectively.

Entrepreneurial Bias

Entrepreneurial bias stems from an individual's previous knowledge and entrepreneurial mindset. This bias is essential for opportunity recognition, as it explains why not everyone identifies the same opportunities. Without entrepreneurial bias, everyone would recognize the same opportunities, leading to a lack of differentiation.

Entrepreneurial bias arises from asymmetry in learning, knowledge, and experiences among people.

Individuals have different learning capabilities, know-how, and backgrounds, leading to varying perspectives on the events they encounter.

These differences mean that while some people see an event as an entrepreneurial opportunity, others may not. Whether active or passive in their approach, every founder experiences entrepreneurship.

This bias shapes how they perceive and act upon potential opportunities, making it a critical factor in the entrepreneurial journey.

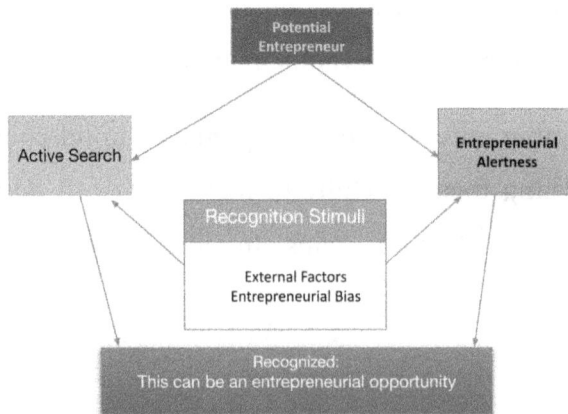

Potential Entrepreneur

Active Search

Entrepreneurial Alertness

Recognition Stimuli

External Factors
Entrepreneurial Bias

Recognized:
This can be an entrepreneurial opportunity

Connecting the Dots

Opportunity recognition is an iterative process in entrepreneurship. Searching may be either passive (entrepreneurial alertness) or active, and events may be recognized as opportunities or not.

The speed and accuracy at which an opportunity is recognized are influenced by external factors and the entrepreneurial bias. However, recognizing an opportunity is just a first step and very far from a final decision.

Opportunity Evaluation

Once an individual recognizes an entrepreneurial opportunity, the next crucial step is evaluating that opportunity.

Evaluation is a vital phase where System 2 of our brain takes over, engaging in rational, slow, and deliberate thinking. This phase requires more time, deeper consideration of various factors, and careful analysis.

Opportunity evaluation is intricately connected to the recognition phase, involving both System 1 and System 2 of our brain. It is a process deeply rooted in dealing with information, making it an exercise in information science.

The individual who has identified a potential opportunity must now make a firm decision, a decision that can be one of the most important in their lifetime.

Evaluating an opportunity is essentially an information game. It involves collecting all necessary information, integrating it with existing knowledge and experience, and ultimately reaching a well-informed conclusion. Founder's opportunity evaluation is not as complicated and analyzed as a VC due diligence process of investment. Most decisions were made by intuition and heart of the founders instead of over analysis. It doesn't mean that there should not be any analysis of evaluation, though.

From Data to Wisdom

The role of information in entrepreneurship is quite evident, as scholars of entrepreneurship and strategy consider it a fundamental principle for decision-making. Before delving into information and its role in opportunity recognition, let's clarify the related terms and how we define them.

DIKW Pyramid

Introduced by Russel Ackoff in 1989, the DIKW pyramid represents a hierarchy for the flow of information and how it transforms into more compressed and valuable elements.

- **Data**: Data consists of raw symbols that convey no inherent meaning but represent distinctions. It results from observations and is unprocessed, having no function, value, or organization until it is made usable in a process.
- **Information**: Information is processed data that is more useful to the audience. It usually answers who, what, where, and when questions.

- **Knowledge**: Knowledge provides instructions and leads to solving problems based on the available information. It usually answers the 'How' questions. The main difference between knowledge and information is efficiency; knowledge is efficient, specified information and is defined as the sum of all known facts.

- **Understanding**: Understanding clarifies the reasons behind presented knowledge or information. It usually answers the 'Why' question. A deeper understanding is also called insight, and some authors have added this to the model.

- **Intelligence**: Intelligence is the ability to specify a problem and set goals to solve that problem by addressing actions based on available knowledge.

- **Wisdom**: Wisdom is an evaluation of understanding and the ability to increase effectiveness.

Information is critical for entrepreneurial opportunity recognition and evaluation. Individuals get their information from various sources, and they process this information differently.

Shane and Venkataraman suggest that people acquire information through "information corridors," but to evaluate opportunities evenly, they must possess similar cognitive abilities. Since individuals have different cognitive abilities, they also evaluate opportunities differently.

Vagheley and Julien introduced two types of information processing methods:

- **Algorithmic Processing**: Deals with patterns, formulas, experience, and intuition to recognize opportunities.
- **Heuristic Processing**: Constructs opportunities using methods such as trial and error, sense-making, interpretation, and intuition.

In both types, the most important factor remains the human factor. The individual's approach to processing information significantly impacts the recognition and evaluation of entrepreneurial opportunities.

Opportunity Trap

Nascent entrepreneurs are particularly vulnerable to falling into opportunity traps. Just as you can't learn to swim without drinking water on your first days, you can't become a successful entrepreneur without making mistakes to gain experience through iteration and trial and error. Many people wish they could go back in time with their current experience and knowledge, but that's not possible. In most cases, when little or no proper analysis is done, opportunity traps are a promising-looking situation that ends with harm. Falling into opportunity traps is just the norm while running your business. It is very normal to face failures and make mistakes, where in most cases, startup founders have gained a lesson out of these sorts of experiences. The learning curve in entrepreneurship is shaped by these traps, helping entrepreneurs refine their decision-making processes and improve their chances of future success.

Opportunity Exploration of The Golden Eagle Entrepreneur

Much like the majestic golden eagle, the golden eagle entrepreneurs employ a blend of sharp vision, accumulated knowledge, and vigilant alertness to explore business opportunities.

They actively seek out potential ventures while maintaining a high level of awareness, allowing them to quickly analyze and respond to emerging trends and gaps in the market.

These entrepreneurs understand the importance of gathering and processing information without falling into the trap of over-analysis.

By balancing instinct with strategic thinking, they efficiently identify and capitalize on opportunities, embodying the resilience and precision of the golden eagle in the entrepreneurial world.

CHAPTER FOUR

FUTURE FORESIGHT

"The future belongs to those who believe in the beauty of their dreams." — **Eleanor Roosevelt**

In the large expanse of the entrepreneurial landscape, visionaries stand as modern-day golden eagles. These individuals, with their keen foresight, soar high above the mundane, spotting opportunities and navigating challenges with an unparalleled perspective.

The future is their domain, and like the golden eagle, they possess a rare combination of sharp vision, strategic insight and defining the future!

On the Essence of Future Foresight

In today's super-fast world, technology is changing really fast. To keep up, we need to be ready for whatever's next. The market is packed, and competition is fierce. That's where entrepreneurs with a disruptive mindset come in - they think outside the box and see what others don't. They're all about future opportunities and creating tomorrow's solutions today.

They don't waste time building something for the present and trying to compete with everyone else. Instead, they innovate and create game-changing products that will shape the future. Having a clear vision of what's to come is key to making it happen- Future foresight. It's all about imagining what the future will be like and building it now.

The Power of Vision

Just like a golden eagle can spot its lunch from miles away, entrepreneurs need to have super sharp eyes on their surroundings. We talked about how to sniff out opportunities and explore them in the last chapter. Now, let's dive deeper into how to keep your eyes peeled for the next big thing.

Think of it like having a sixth sense for what's happening around you. You need to be aware of the latest trends, what people are talking about, and what's missing in the market. It's like having a radar that's always on the lookout for the next opportunity. And when you spot it, you can swoop in like an eagle and grab it!

Visionary entrepreneurs have the capacity to look beyond the immediate circumstances and see the bigger picture. They think ahead by visualizing the potential impact of their ideas and foresee how their ventures could transform industries or even society as a whole.

The vision is not only leading them to the futuristics approach, but also inspires and motivates the team. Visionary entrepreneurs have a great sense of imagination and through their dreams, they make the future themselves.

On Forecasting the Future

Humans have always been curious about what's coming next. We try to predict the future in all sorts of ways, from checking the weather app on our phones to trying to guess what's going to be trending on social media.

But when it comes to business, forecasting the future is a whole different ball game. It's not just about being curious, it's about survival. Making the wrong predictions can be disastrous for a company, while making the right ones can be a total game-changer.

So, how do we do it? How do we gaze into the crystal ball and figure out what's coming next? Well, that's what we're going to explore.

There are a variety of future forecasting methods that enable us to predict the future to anticipate changes and shape our strategic direction. These methods provide a structured approach to understanding potential futures and prepare for them. There are three major types of methods to predict the future in business: Qualitative method, quantitative method, and the hybrid method.

Qualitative Methods of Forecasting the Future

Scenario Planning

Scenario planning is a strategic method designed to explore and prepare for multiple potential future environments. Unlike traditional forecasting, which attempts to predict a single future outcome based on current trends, scenario planning acknowledges the uncertainty and complexity of the future by creating a set of diverse, plausible scenarios.

Each scenario represents a different way that the future might unfold, taking into account various key drivers and uncertainties.

This approach involves the exploration of multiple futures rather than predicting one specific outcome, focusing on key uncertainties that could significantly impact the future.

These uncertainties form the basis for developing different scenarios, allowing organizations to develop flexible strategies that are robust across various potential futures and enhance their adaptability.

Scenario planning typically uses detailed narratives to describe how different factors might interact to shape the future, helping stakeholders visualize and understand the implications of each scenario.

This method aids decision-making by highlighting potential risks and opportunities, enabling organizations to make more informed strategic choices.

It finds applications in business strategy, where it helps companies anticipate market changes, competitive dynamics, and technological advancements; in public policy, assisting governments in preparing for various socio-economic and environmental challenges; in environmental planning, supporting the planning for different environmental futures such as climate change scenarios; and in innovation and technology, guiding research and development efforts by considering future technological and market trends.

The scenarios could be split into, most probable, favorable, and unfavorable conditions that might happen in the future and based on those scenarios we can just be more prepared. Scenario planning helps us build a sort of contingency plan for the future and face the uncertainty.

Scenario planning is categorized as the qualitative method of future forecasting, but the sensitivity analysis in financial projections is also sort of scenario planning, where you see the impact of different future scenarios on your projected business. For example, if your revenue drops 5% or costs increase 10% what will be the impact of those events on your cash flow, P&L, or even your startup valuation?

Delphi Method

The Delphi method is a structured communication technique used to forecast the future or make informed decisions based on the opinions and judgments of a panel of experts.

It is widely used in various fields such as business, healthcare, and technology for forecasting and decision-making. The process typically as follows:

First, a group of experts is selected based on their knowledge and expertise in the relevant field. These experts are then asked to answer a series of questions or provide their opinions on a specific issue or future event through a questionnaire designed to gather a broad range of views.

Responses are anonymized to prevent influential persons from dominating and to ensure unbiased feedback. The responses are then summarized, and the summary is sent back to the experts. These summaries include syntheses of general agreement, although they also highlight areas of disagreement.

The summaries are then given back to the experts so they can provide additional rounds of feedback or revised opinions. Usually, two to four rounds of iteration are conducted, refining answers toward convergence to a consensus or a more accurate forecast.

Finally, a concluding analysis summarizes the experts' opinions, representing their collective judgment and forming the basis for forecasting or decision-making.

Key features of the Delphi technique are as follows:

- It is an anonymous process, so responses are not influenced by a few strong personalities, thereby avoiding groupthink.

- It is iterative, allowing participants to re-evaluate their opinions based on feedback.
- It provides controlled feedback, facilitating understanding of different perspectives and moving toward consensus; and it is statistically aggregated, where the most frequent response likely represents a statistical summary of the group's views, including measures of central tendency and dispersion.

The Delphi method has been applied to technology forecasting, policy making, business strategy, healthcare, and environmental studies.

The basic logic of the Delphi approach is to draw upon pooled expert knowledge to make forecasts or decisions that are more reliable and valid than those developed by any single individual or unstructured group.

Backcasting

Backcasting is one of the strategic planning tools that set out from the definition of a desirable state or goal in the future, detailing step-by-step the strategies needed to achieve a given future.

It is the approach of moving backward from attainment of a certain end result. This is a future-oriented method of trying to get to a desired state rather than predicting what is likely to transpire because of the existing trend.

It uses the problem-solving approach whereby there is identification of the gap between the current situation and the desired future, then seeking ways to bridge it.

This allows backcasting to be a holistic view, which has considered many factors and influences that can affect achievement of a desired future and allows identification of multiple pathways to accommodate different strategies and actions.

The process of backcasting has several steps, the first being starting with clearly articulating the long-term goal or vision, ensuring it is SMART, that is *specific, measurable, achievable, relevant,* and *time-bound.*

It does assess the current state in relation to the desired future by identifying strengths, weaknesses, opportunities, and threats of the key factors, stakeholders, and systems involved.

It clearly identifies the gap between the current situation and the way it ought to be in the future, understanding the barriers and problems needed to be overcome.

This is then followed by brainstorming and an exploration of alternative ways to bridge that gap with innovative solutions, policies, and technologies, along with actions. Consider the feasibility and possible impact of every strategy and detail an action plan on the implementation of selected strategies that include steps towards the achievement of short- and intermediate-term goals, resources to be utilized, responsibilities involved, timelines to be followed, and milestones set to determine completion.

The plan is put into motion and changes are constantly made as per the feedback and evolving circumstances.

Last but not least, regular reviews on the progress and effectiveness of those actions are carried out, reflecting upon what has been learned so far and noting the necessary adjustments in order to stay on track toward that goal.

This structured approach in backcasting serves great latitude in enabling long-term targets to be attained while still providing flexibility to encourage innovative solutions. Backcasting is one of the top-notch corporate strategy planning methods that should encompass both long-term and short-term objectives of the company.

The reason for the author's preference for backcasting in strategic planning is because it involves team members in most brainstorming sessions, and further inculcates the culture of future foresight and innovation into the company. This is a great tool for ideation, corporate strategy planning, contingency planning, and even personal development.

Future Wheel

The Future Wheel is a visual brainstorming tool used to explore the potential direct and indirect consequences of a particular change or event.

Developed by futurist Jerome C. Glenn in 1971, it helps individuals and organizations think systematically about the ripple effects of a new trend, innovation, or decision.

By mapping out these consequences, the future wheel can aid in understanding the broader implications and preparing for different scenarios.

The future wheel has several key features that make it a valuable tool for strategic planning and decision-making. By offering a structured approach, it encourages systematic thinking by breaking down complex changes into manageable parts.

One of its significant benefits is identifying ripple effects, not only recognizing immediate (first-order) consequences but also subsequent second-order and third-order effects.

The first step to building the Future Wheel is to put something central around it; a certain change or event that takes place. It can be a new technology, a change in market trend, decision on policy, or anything else of huge proportion.

Brainstorm the first-order consequences by drawing lines from the central in a way that symbolizes the immediate, direct effect of the central change, and write these consequences on paper spread around the central change.

Then, even more specifically, identify second-order consequences by brainstorming the likely effects of those secondary effects and draw lines out from each first-order consequence to represent it, writing out branches from each first-order effect.

This can be continued to third- and higher-order consequences if necessary, in exploring the ripple effects even farther away from the central change.

Once complete, analyze and interpret the future wheel to pick out patterns, opportunities, risks, and potential strategies. This amounts to looking into the wheel and including everything interconnected: it means an understanding of how various conjoined consequences interact with each other.

Thorough review should therefore amount to building a panoramic understanding on future possibilities of outcomes and be helpful in strategic decision-making.

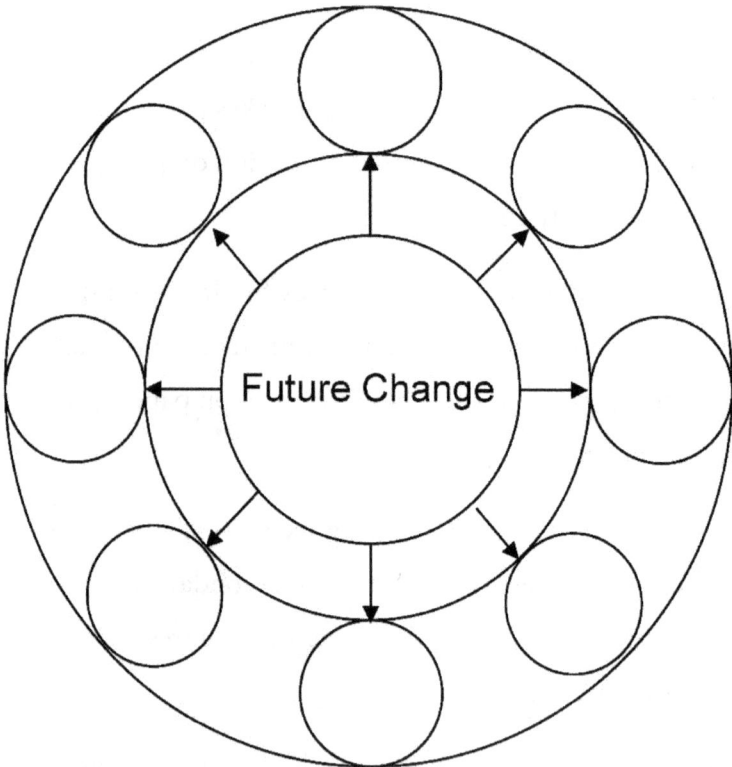

Although there are some recommendations from futuristics to use SWOT and PESTLE analysis as future prediction tools, we do believe that these are more fit tools for the short to mid-term analysis and there is not much degree of futuristic approach in those models.

Quantitative Future Prediction Methods

Quantitative future prediction methods use numerical data and statistical techniques to forecast the future key trends and key numbers for the business. These methods include traditional statistics methods along with the new-rising predictive AI and serve as the prediction tool in the business.

Time Series Analysis

Time series analysis involves the study of data points collected or recorded at specific time slots to identify patterns, trends, and seasonal changes.

This analytical method helps us to understand the data over time and forecast the future based on identified patterns of historical data. Several methods are employed in time series analysis to achieve accurate and reliable predictions.

One common method is moving averages, which smooths out short-term fluctuations to highlight longer-term trends, making it easier to see the overall direction of the data.

Exponential smoothing is another technique that applies weighted averages to past observations, with more weight given to recent data, allowing the model to react more quickly to changes.

ARIMA (Auto Regressive Integrated Moving Average) is a sophisticated approach that combines autoregressive and moving average models to predict future points in a series, accounting for various lags and differences in the data.

Seasonal Decomposition of Time Series (STL) is used to decompose a time series into seasonal, trend, and residual components, providing a clear view of each component's contribution to the overall data pattern.

Time series analysis has numerous applications across different fields. In sales forecasting, it helps businesses predict future sales based on past performance, allowing for better inventory management and marketing strategies.

In finance, it is used to predict stock prices and market trends, aiding investors in making informed decisions. Weather forecasting relies heavily on time series analysis to predict future weather conditions based on historical weather data, contributing to better preparation for adverse weather events.

Through these applications, time series analysis proves to be a crucial tool in various industries, offering insights and predictions that drive strategic decisions and planning.

Regression Analysis

Regression analysis is a statistical method that examines the relationship between dependent and independent variables to predict future values.

By understanding and modeling these relationships, regression analysis helps in making informed predictions and decisions based on data trends and patterns.

One of the primary methods used in regression analysis is linear regression, which models the relationship between two variables by fitting a linear equation to the observed data.

This method assumes that changes in the dependent variable can be explained by changes in the independent variable. Linear regression is straightforward and widely used for its simplicity and interpretability.

Multiple regression extends linear regression by including multiple independent variables in the model. This approach allows for a more comprehensive analysis by considering the influence of several factors on the dependent variable. By incorporating multiple variables, multiple regression can provide more accurate and detailed predictions, making it a powerful tool for complex datasets.

Polynomial regression, on the other hand, models the relationship using polynomial equations. This method is particularly useful when the relationship between the variables is not linear, allowing for the modeling of more complex, curvilinear patterns in the data. Polynomial regression can capture the nuances of data trends that linear models might miss, providing a better fit for certain types of data.

Logistic regression is used for binary outcome prediction, such as yes/no or success/failure scenarios. Unlike linear regression, logistic regression models the probability of a binary outcome based on one or more predictor variables.

This method is essential in fields where predicting categorical outcomes is crucial, such as medical diagnosis, credit scoring, and marketing response modeling.

Machine Learning Algorithms

Machine learning algorithms use statistical techniques to learn from data and make predictions. These algorithms can be categorized into three main types: supervised learning, unsupervised learning, and reinforcement learning.

Supervised learning algorithms learn from labeled data, meaning the input data comes with corresponding output labels.

Common supervised learning algorithms include linear regression, which models the relationship between variables; decision trees, which split data into branches based on input features; random forests, an ensemble method that combines multiple decision trees to improve accuracy; and support vector machines (SVM), which find the optimal hyperplane to separate different classes in the data.

Unsupervised learning algorithms identify patterns in unlabeled data, where the algorithm must find the structure in the input data on its own. Examples of unsupervised learning include k-means clustering, which groups data points into clusters based on similarity, and principal component analysis (PCA), which reduces the dimensionality of data while preserving its variance.

Reinforcement learning involves algorithms learning optimal actions through trial and error, where the algorithm receives feedback in the form of rewards or penalties. Q-learning is a popular reinforcement learning algorithm that teaches an agent to act optimally in an environment to maximize cumulative rewards.

Machine learning algorithms have a wide range of applications. Personalized recommendations, such as those used by streaming services or online stores, rely on machine learning to suggest products or content based on user behavior.

Fraud detection systems use machine learning to identify unusual patterns that may indicate fraudulent activity. Predictive maintenance leverages machine learning to anticipate equipment failures before they happen, allowing for proactive maintenance and reducing downtime.

Predictive AI, a popular AI-powered quantitative method, refers to using artificial intelligence (AI) and machine learning (ML) techniques to analyze historical data, identify patterns, and make predictions about future events or behaviors.

Predictive AI leverages algorithms that learn from data to forecast outcomes, enabling businesses and organizations to make data-driven decisions and optimize their operations.

Key components of predictive AI include data collection and preprocessing, where relevant historical data is gathered and prepared for analysis by cleaning, normalizing, and transforming it.

Feature engineering involves selecting and creating features (input variables) most relevant to the prediction task. Model selection is about choosing appropriate machine learning algorithms based on the nature of the data and the prediction problem.

Training and validation entail training the model on historical data and validating its performance using metrics such as accuracy, precision, recall, and F1 score.

Prediction and deployment involve using the trained model to make predictions on new data and deploying the model into production for real-time or batch predictions. Finally, monitoring and updating involve continuously monitoring the model's performance and updating it with new data to maintain its accuracy and relevance.

Econometric Models

Econometric models are sophisticated analytical tools that combine economic theory with mathematics and statistical inference to measure economic phenomena and forecast future trends.

These models help economists and policy analysts understand complex economic relationships, enabling them to make informed decisions based on empirical data.

One key method in econometric modeling is Vector AutoRegression (VAR). The main advantage of VAR models is that they account for interdependencies among multiple time series, allowing for the investigation of how different economic variables influence each other over time.

VAR models capture the dynamic dependence among the included variables and provide information on causal relationships or the impact of economic shocks throughout the system.

Cointegration models are another technique used in econometrics to identify long-term relationships between non-stationary time series. Cointegration indicates stable long-run equilibrium relationships among individual time series trends, even if each series may wander randomly. This technique is extremely useful for studying variables like interest rates and inflation or investment, which tend to move together in the long run.

Error correction models (ECMs), based on cointegration, model deviations of short-term dynamics from long-term equilibria that develop due to the lag structure of the disequilibrium correction function.

ECMs are designed to correct any short-term deviations, ensuring that the variables move toward equilibrium in the long run. This approach provides a sophisticated understanding of the relationship between variables, accounting for both short-run fluctuations and long-term trends.

In financial markets, econometric models offer insights into trends, asset prices, and investment risks, supporting investors in making informed choices. By integrating economic theory with mathematical and statistical techniques, econometric models provide powerful tools for analyzing and forecasting economic phenomena.

They account for the intricate workings of the economy, offering valuable guidance for economic planning and policy.

Simulation Models

Simulation models employ computational techniques to imitate real-world processes, providing a valuable means to assess the impact of different variables on complex systems.

By creating a virtual environment that represents the behavior of a system, these models enable analysts to experiment with various scenarios and predict potential outcomes without interfering with the actual system.

One of the most widely used methods in simulation modeling is Monte Carlo Simulation. This technique utilizes random sampling and statistical modeling to estimate the probability distribution of outcomes.

By repeatedly running simulations with varying input values, Monte Carlo Simulation can provide a comprehensive picture of the possible results and their associated probabilities. This method is particularly effective in situations involving uncertainty and risk, as it allows for the evaluation of a wide range of scenarios and the quantification of potential risks and rewards.

Discrete Event Simulation is another method, which models the operation of a system as a sequence of events in time. Each event occurs at a specific point in time and changes the state of the system.

Discrete Event Simulation is particularly useful for systems where changes happen at discrete intervals, such as manufacturing processes, logistics operations, and customer service systems. By modeling these events and their interactions, this method helps in identifying bottlenecks, optimizing resource allocation, and improving overall system efficiency.

System Dynamics is a method that focuses on modeling the behavior of complex systems over time using stocks, flows, and feedback loops. This approach is particularly suited for understanding how different components of a system interact and evolve over time.

System Dynamics models are often used to study long-term strategic issues, such as population growth, economic development, and environmental sustainability. By capturing the feedback loops and time delays inherent in complex systems, this method helps in identifying leverage points and designing policies that lead to sustainable outcomes.

Simulation models can be used in risk analysis to evaluate the potential impact of uncertain future events, such as financial market fluctuations, natural disasters, and operational risks. By simulating different risk scenarios, organizations can develop strategies to mitigate potential losses and enhance resilience.

In supply chain management, simulation models help in optimizing inventory levels, production schedules, and distribution networks, leading to improved efficiency and cost savings.

Project management also benefits from simulation models, as they enable the assessment of project timelines, resource requirements, and potential delays, facilitating better planning and execution.

Hybrid Methods

Judgmental Forecasting

Judgmental forecasting is a forecasting technique that involves developing future scenarios or future trends of events based on expert judgment, intuition, and subjective estimates—not only on quantifiable data and statistical models.

This is particularly helpful when the environment for future change is expected to be very different from the past or when historical data are limited or of poor quality. Judgmental forecasting, either alone or combined with other quantitative methods, enhances the accuracy and relevance of the prediction.

A few salient features of judgmental forecasting include its dependence on expert opinion, where the experience and insight of the expert add the needed background knowledge to the model. It includes subjective judgment, which is very useful in situations dealing with uncertainty or ambiguity.

The approach is flexible and can accommodate new information, changed circumstances, and unique contexts that quantitative models may not fully grasp. It also uses qualitative data—market insights, industry trends, and anecdotal evidence—things difficult to convert into numbers.

These forms of techniques are used in judgmental forecasting: The Delphi method and scenario planning can also be integrated in the judgmental forecasting. Intuitive judgment relies on the intuitive insights and gut feelings of knowledgeable people to come up with a forecast, especially when data are scarce or time is of the essence.

As mentioned, the intuition of entrepreneurs plays an important role in their decision making. Panel consensus entails a panel of experts deliberating and deciding on the expected outcome, which can take place during in-person meetings, virtual discussions, or structured workshops. Historical analogy applies past experiences and analogies to parallel situations to predict the future.

Applications of judgmental forecasting are helping to forecast the success and demand for new products or services with limited or no historical information, relying on expert opinions and market insights to evaluate potential adoption rates, customer preferences, and competitor dynamics.

Judgmental forecasting is significant in crisis management, making forecasts and decisions during crises such as natural disasters, economic downturns, or geopolitical events. In such scenarios, expert judgment and scenario planning help with risk preparation, contingency planning, and timely decision-making.

Technological forecasting involves anticipating new technologies and innovations and their impacts; through the Delphi method and expert panels, invaluable insights into technological trends, adoption rates, and potential disruptions are drawn.

It supports public policy decisions, forecasting social, economic, and environmental impacts using scenario planning and expert consensus to help policymakers evaluate different policy options and their potential outcomes.

Composite Forecasting

Composite forecasting combines multiple forecasting methods to produce a single, comprehensive forecast, leveraging the strengths and mitigating the weaknesses of individual methods to achieve more accurate and reliable predictions.

By integrating a mix of quantitative and qualitative techniques, composite forecasting provides a holistic view of future trends.

Key features of composite forecasting include its ability to combine multiple methods, such as time series analysis, regression models, machine learning, and expert judgment.

This integration enhances accuracy by averaging or blending different forecasts, reducing the impact of errors and biases inherent in individual methods.

It also offers flexibility by including diverse data sources and perspectives, making it adaptable to various forecasting scenarios. Additionally, composite forecasting mitigates risk by diversifying the forecasting approach, thus spreading the risk of inaccuracies across multiple methods.

The steps in composite forecasting begin with selecting the forecasting methods that are suitable for the specific scenario. Once the methods are chosen, each is applied to the available data to generate individual forecasts.

These individual forecasts are then combined using statistical techniques such as averaging, weighted averaging, or more complex models like ensemble methods.

Finally, the composite forecast is validated by comparing it with actual outcomes, allowing for necessary adjustments to ensure its accuracy.

Foreseeing the future is a vital part of running a business in particular as an entrepreneur, but the golden eagle entrepreneur chooses another option. Forecasting the future is still a reactive approach to mitigate risks and plan for contingencies, while there is a creative approach of dealing with the future.

"The best way to predict the future is to create it." – Abraham Lincoln

Building the Future

The golden eagle entrepreneur never stays reactive to respond to events and conditions. Although it is wise to have a future prediction and an educated guess, the golden eagle entrepreneur takes a hands-on approach to disrupt all predictions and trends through innovation.

The out-of-the-box mindset drives them to take action in the pursuit of their dreams and eventually build the future.

One of the traps that kills the entrepreneurial spirit is market research. Here is the story from the author:

"I have been advising a Corporate Venture Capital (CVC) during its establishment, where my work as their advisor took only 6 months. The company has been one of the most successful firms with lucrative sales of home appliances and revenue. They wanted to create a venture capital arm with the aim of getting into some new industries in order to invest their money in some even more lucrative industries.

We used to scout startups, take them into the investment funnel, do the screening, and present in the the investment committee and eventually get a rejection on the deal from the committee.

After months of taking startups into that investment committee I realized the blocker: "Market research that was part of the document was the most important decision making criteria for the investment committee in which I had one vote and my vote and arguments were not persuasive enough for investment.

The investment committee was the same as the board members and they got used to making money on every initiative from day 1 for years. Such a mindset does not allow them to embrace failure or take risks to build the unknown future and hence, there was no future mindset in that company. After 6 months, I left the company and CVC was shut down after a year."

Market research and prediction tools rely on the present and short-term future and even if they produce some results for the longer-term future, it is about something that exists and already is working.

Most market insight reports are on the current technologies that exist and are already there. There are many founders these days that are working on AI and there are too many events and investments rushing towards that field. However, at the time that you are reading this page, some golden eagle entrepreneurs who imagined a future are working to bring a new technology into existence; a technology that is not there yet.

The pitfall of leading the investment flows toward trending segments and technologies is that sometimes they create a bubble and create lots of opportunity traps. The Dotcom bubble is an example; while many investors rushed their money to dotcom, some invested in the future technology.

Building the unknown future is about pushing boundaries and redefining what is possible. It is about daring to be ambitious enough to dream big and acting on those dreams with purpose and determination.

By navigating the matrix of known and unknown described in chapter ten, innovators can chart a course through uncertainty, uncovering new opportunities and creating a future that is not only envisioned but realized.

The same as the golden eagle, entrepreneurs tend to build the future by assuming that the past is not enough for a better life and better impact. This is one part of the brain and the mindset of the golden eagle entrepreneur to dream about the future and take an innovative approach in building the future.

CHAPTER FIVE

INNOVATION

"Don't be satisfied with stories, or how things have gone with others. Unfold your own myth." — **Rumi**

The golden eagle stands at the pinnacle of innovation, so do the golden entrepreneurs whose groundbreaking ideas and inventive strategies build the future!

Not satisfied with the status quo, they have the ability to identify and harness future technologies that reshape the industries and set new standards for excellence.

With a relentless drive for innovation, the golden eagle entrepreneurs continually push the boundaries of what is possible, inspiring a new generation of entrepreneurs to dream bigger and achieve more.

An Overview on Innovation

Innovation is more than just new products; it is the creation of better ideas, methods, service, or solutions that is derived from real value in bringing positive change.

Innovation translates creative concepts into real-world outcomes that increase efficiency and effectiveness or meet unfulfilled needs. In business terms, it's the ability of an organization to conceive, design, develop, commercialize, and scale offerings in an ongoing and sustainable manner with significant value addition.

Innovation is a complex and multifaceted concept that goes way beyond mere uniqueness or invention. Innovation represents the whole process of transformation of creative ideas and inventions into valuable goods, services, or processes capable of changing society, the economy, or some particular sphere.

Invention can also lead to simplification of a product by subtracting certain features. A product can be enhanced by making it simple, such as the removal of buttons on a website that impact user experience. By simplifying numerous things, one improves user experience and therefore a product is utilized and attractive to potential customers.

There have been several great innovations where products have been made simple in order to improve their use, a concept valued by many consumers. Market innovation involves the incorporation of a set of different product features into a single offering. This is particularly effective where one can ride the success of a given product to make others better.

This is achievable through understanding why a successful product attracts customers: such insights often come from buyer feedback, online reviews, and insights from sales reports, which might point to some key features responsible for driving such success.

Such information is vital to ensure that the product development in any business is done and sustained both now and into the future. A business would be placed at a vantage point of detecting what needs to be improved upon by critically reviewing what they offer and looking for improvement, which would make it outstanding from the rest. This can be doing an add-on with a few new features or improving one's experience or ease of doing it.

Innovation is simply not the production of something that did not exist before, and it takes many forms, such as financial, cultural, and practical and social value.

Creativity, then, is innovation linked to utility and enterprise, and an innovative product "creates value for others." Whereas the invention needs something to be made different, innovation is that of developing a desirable and usable product or service for the customer to consume.

A company must transform a creative idea into an innovative product with significant short- or long-term value. The psychological method of converting a concept into creativity is going through the creative process and requires strategic preparation for the production and marketing of the product.

On the Essence of Innovation

Entrepreneurship and innovation are inherent to one another; entrepreneurship is about imagining the world not as it is, but as it could or it will be.

This visionary aspect compels the entrepreneur to shake up the status quo, push boundaries, and develop solutions that were once inconceivable.

As mentioned earlier, building the future is likely only possible through innovation, which serves as a competitive advantage for companies and startups. Innovation becomes the cornerstone for new social impacts, value generation, and improved quality of life. Innovation in entrepreneurship can manifest through various characteristics. It can be seen in groundbreaking products that revolutionize industries, novel business models that disrupt traditional practices, and unique services that redefine customer experiences.

The essence of innovation lies in its ability to transform ideas into tangible realities, fostering an environment where continuous improvement and adaptation are essential. For entrepreneurs, the journey of innovation is both challenging and exhilarating. It demands a blend of creativity, resilience, and strategic thinking. An entrepreneur must be ready to find unmet needs and market gaps, many times using the latest technology and data insights to craft a solution that can not only be viable but also scalable.

Relentless innovation requires one to risk or embrace failure as another learning curve because each setback brings an individual closer to breakthrough realization.

Moreover, the impact of entrepreneurial innovation extends far beyond individual success. Innovative initiatives create employment and diversify economies, enhancing the competitiveness of nations in the global market. They stimulate technological development and research, leading to industrialization.

Therefore, the spirit of innovation is a key driving force behind economic vitality and societal welfare. The fast pace of change in today's world heightens the need for a culture of innovation within the entrepreneurial ecosystem. This culture is nurtured and supported by governments, educational institutions, and private-sector entities, providing resources, mentorship, and funding for ambitious young entrepreneurs to carry out their innovative ideas.

Close collaboration and creating an enabling environment are essential to maintaining the speed of innovation and ensuring it remains a powerful force for positive change.

Types of Innovation

Innovation in Product/Service

This is the most well-known type of innovation. Product or service innovation introduces a completely new product or service to the market, targeting a gap in the market.

At a more disruptive level, it can create new demand and new markets or disrupt existing ones. Examples include electric cars, the introduction of smartphones, and the rise of two-sided platforms like Uber.

Innovation in Process

Innovation in processes and operations occurs when a new change in current processes happens. This mostly leads to improved efficiency and cost reductions, but in some cases might increase value and impact of the company.

Toyota's lean manufacturing is one remarkable example of innovation in a process that revolutionized the automotive industry by reducing waste and improving production efficiency.

Innovation in Paradigm

Innovation in a paradigm goes beyond incremental changes; it represents a shift that largely defines new industries, new markets, and ultimately new standards.

Such disruptive breakthroughs change the basis of competition in existing industries, creating opportunities for competitive advantage, improving customer value, and contributing to societal progress.. Embracing paradigm innovation requires vision, resilience, and a commitment to moving ahead in uncertainty, but the rewards are deep—paving the way toward a brighter, more sustainable future.

The Extent of Innovation

Incremental Innovation

Incremental innovation can be described as the process by which a series of small, ongoing improvements in products, services, or processes are made. It enhances what already exists instead of developing something entirely new.

This type of innovation often comes in terms of updates, optimizations, and improvements that might add up to increased value in an existing product or service. Companies improve effectiveness, reduce costs, and meet customer needs by leveraging current resources and knowledge.

Incremental innovation is low-risk in its nature and leads to continuous, consistent growth. This helps the organization adapt to market changes without suffering major disruptions. Incremental innovation is very essential for the survival and progression of existing products, services, and processes.

Although the impact is too fundamental, incremental innovation still helps businesses keep their competitiveness intact, enabling customer satisfaction and ensuring sustained growth through continuous, manageable improvements.

As an example of incremental innovation we see Apple regularly releases new versions of the iPhone with incremental improvements such as better cameras, faster processors, and enhanced software features. Each new model builds upon the previous one, adding new functionalities and refining existing ones without drastically changing the core product.

Breakthrough (Radical) Innovation

Radical innovation, in contrast, introduces groundbreaking changes that significantly alter the way industries operate. This type of innovation involves the development of entirely new products, services, or technologies that can create new markets or disrupt existing ones.

Radical innovation requires substantial investment in research and development and often carries high risks due to the uncertainty involved. However, the potential rewards are equally high, as successful radical innovations can lead to significant competitive advantages and reshape entire industries.

Tim Berners-Lee's creation of the World Wide Web in 1989 introduced a completely new way of accessing and sharing information. This radical innovation transformed communication, commerce, and entertainment, creating an entirely new digital landscape and profoundly impacting society and the global economy.

Architectural Innovation

Architectural innovation involves reconfiguring existing technologies and components to create new products or systems.

While the core components remain largely unchanged, their arrangement and interactions are altered to offer new functionalities or improved performance. This type of innovation often leverages existing technological capabilities in novel ways, opening up new applications and markets.

Architectural innovation requires a deep understanding of how different elements can be recombined to produce innovative outcomes. As an example, cloud computing reconfigures existing technologies (servers, storage, networking) to provide scalable and flexible computing resources over the internet. This architectural shift allows businesses to access and manage resources more efficiently and cost-effectively, using a new configuration of existing components.

Disruptive Innovation

Disruptive innovation refers to innovations that initially target a niche market or under-served segment but eventually displace established products or services.

These innovations start as lower-cost alternatives that offer simpler or more convenient solutions compared to existing offerings. Over time, they improve in quality and performance, attracting a broader customer base and challenging the dominance of established players.

Disruptive innovations often democratize access to products or services, making them more affordable and accessible. Netflix began as a DVD rental service but disrupted the traditional video rental industry by introducing streaming services.

This initially served a niche market but eventually revolutionized how people consume media, leading to the decline of physical rental stores like Blockbuster.

Type of Innovation	Cost	Degree of Innovation	Risk	Improvements
Incremental Innovation	Low to Moderate	Low to Moderate	Low	Enhancements to existing products, processes, or services
Architectural Innovation	Moderate to High	Moderate to High	Moderate	Reconfiguration of existing technologies for new applications
Disruptive Innovation	Initially Low, increases over time	High (Over Time)	Moderate to High	Initially targets niche markets, evolves to disrupt industry
Radical Innovation	High	High	High	Creation of entirely new products, technologies, or markets

The table shows the innovation types and costs, risks, and innovation involved.

Open Innovation

Open Innovation refers to the departure from a classical innovation process of in-house innovation development such as internal R&D.

Open Innovation describes the innovation process as a multifaceted, open search and solution process that occurs between multiple actors beyond company boundaries.

This opening of the innovation process to external input and the outsourcing of tasks to actors who have special competencies or local knowledge for their solution creates new potentials.

For example, a manufacturing company can improve a variety of key performance indicators of the innovation process through Open Innovation. A generic classification in this regard distinguishes between:

- **Time-to-Market**: Shortening the period from the start of product development to its market launch.
- **Cost-to-Market**: Reduction of the actual costs incurred and attributable to the product from the beginning of its planning to its market launch.
- **Fit-to-Market**: Increasing the market acceptance of a new product in terms of a positive purchasing attitude of the consumers (thus creating a higher willingness to pay).
- **New-to-Market**: Increasing the perceived novelty of an innovation by consumers, thereby enhancing the attractiveness of the corresponding product.

Open innovation is a wide and complex concept that helps larger enterprises facilitate innovation within their firms. Some of the most popular forms of open innovation are as follows:

- **Hackathons**: Organizing competitions where entrepreneurs can develop solutions in a short period of time.
- **Innovation Jams**: It is about intensive brainstorming sessions that help generate a competitive environment for new ideas.

- **Co-Creation Work:** Innovative entrepreneurs will bring up new products in a collaborative way.
- **Innovation Labs:** The resources and tools to external investors and startups can be offered here to develop new product concepts.
- **Innovation Competitions**: Aim to generate input for the early phases of the innovation process and encourage innovative ideas through competition among various users.

- **Corporate Venture Capital (CVC):** It is defined as direct investment in start-ups by established companies through their in-house dedicated venture capital arms.

  This kind of engagement aims to access the required financial resources for startup's growth, at the same time providing the corporation insight into new technologies and trends within its market environment.

  A good example could be Google Ventures or Intel Capital, which have invested hugely in numerous start-ups to stay ahead in technology and innovation. Such investments create the potential for high financial returns and, at the same time, create strategic partnerships to drive long-term growth.

- **Joint Ventures and Partnerships:** There will be co-operative works between corporations and startups to achieve common business goals.

  This collaboration would allow both parties to combine their strengths and resources. For example, BMW partnered with the startup ChargePoint for the development of electric vehicle charging infrastructure.

In those ways, companies are able to move into new markets and technologies while sharing their risks and rewards with their start-up partners.

- **Corporate Spin-Offs and Spin-Ins:** Corporate spin-offs involve creating new independent companies from internal projects (Intrapreneurship) or divisions that no longer align with the parent company's core strategy.

  However, Spin-ins, refers to integrating external startups or ventures into the corporation. An example of a spin-off is PayPal's separation from eBay, which allowed both entities to focus on their core businesses. Spin-ins can be seen in Cisco's acquisition strategy, where it integrates smaller tech firms to enhance its technological capabilities and product offerings.

- **Accelerators and Incubators:** These provide most early-stage startups with funding, mentorship, and office spaces. Normally, they work with a fixed-term cohort of startups to accelerate business operations.

  The Barclays Accelerator, powered by Techstars, provides in-depth mentorship to fintech startups with an opportunity to access the Barclays network.

  Incubators may give startups more extended support toward the development of their business models. One example is AT&T's

Aspire Accelerator, which provides a case of how corporations can foster startups into the fold and provide them with a supportive environment to foster innovation and drive growth.

Examples of successful open innovation include Procter & Gamble's Connect + Develop program, which sources new product ideas externally, and LEGO's LEGO Ideas platform, where fans submit and vote on new product ideas that are developed into official sets.

General Electric's Ecomagination Challenge invited external innovators to submit ideas for clean energy solutions, helping GE identify and invest in promising technologies. NASA's use of crowdsourcing platforms like InnoCentive to solve complex scientific problems illustrates how engaging with external innovators can access a vast pool of knowledge and creativity.

The Rise of Corporate Startup Engagement (CSE)

Corporate startup engagement is a strategic approach where larger corporations collaborate with startups to drive innovation, enter new markets, and solve complex problems.

This engagement can take various forms, each offering unique benefits and opportunities for both the corporation and the startups. By using the agility, creativity, and cutting-edge technologies of startups, corporations can improve their innovation capabilities and stay competitive.

Benefits of Corporate Startup Engagement

Engaging with startups offers multiple benefits for corporations. It accelerates innovation by bringing fresh ideas and technologies into the company, as seen with Google Ventures' diverse portfolio of investments.

These collaborations also enable corporations to enter new markets and explore new business models, illustrated by BMW's joint ventures in electric mobility.

Additionally, corporate startup engagement fosters a culture of innovation and agility within established companies, promoting creativity and risk-taking. Successful partnerships and investments can yield significant financial returns, further enhancing the corporation's value.

The author has extensive experience in open innovation mostly by launching CVC and corporate incubators/accelerators and in the last one when he established and led a corporate innovation center and investment arm for a large e-commerce platform:

"We turned a factory hall into a startup factory where we only focused on creating AI solutions with the help of startups. There are several reasons why corporations do not innovate like startups. When you grow, your company loses the agility, this is inevitable.

You also lose the entrepreneurial spirit when everyone working in the company is payroll based and after some times many employees put his working life on auto-pilot. Besides that, larger companies already have a backlog of pending tasks and products to develop and this does not allow them to focus on innovation. There are challenges as well.

In implementation of a successful CSE program, you will face NIH syndrome, which is a high resistance between your internal tech team and the startup. They will block the startup or any other solutions that are Not Invented Here.

This syndrome is the killer in corporations when it comes to their open innovation practices. There are other challenges too. Corporates that earn money well, are not patient enough for long-term innovations. CSE requires patience and an innovative mindset. It requires a golden eagle intelligence!"

Innovation Mindset

The golden eagle entrepreneurs have an innovative mindset to change the future. No matter where they are, they build the future by innovating. Innovation and creativity are also at the heart of the Golden Eagle Entrepreneurs' success. They are not satisfied with the status quo, always seeking to disrupt and redefine industries.

Their innovative spirit is fueled by a deep understanding of market needs and an ability to anticipate the future, enabling them to implement creative solutions effectively. They have an active problem solver and innovation builder brain.

As the author puts it:

"People come to me and ask about how to be innovative and how to train their brain to become more innovative. I always ask them to make

one important assumption: **Everything we do and every product we use right now are the worst ones in the world. Define and build the best now.**

Innovation starts with finding the problem; once the problem is identified, we are halfway there. Loving old solutions are the killers of innovation and that's why a small percentage of people possess an innovative mindset. When we see an innovation, we think that this is the extent of it, until we see the next one disrupting it."

This way of thinking and mindset seems to be easy, but it is one of the hardest things to do in the world. We think that we have great technologies now, because they are working perfectly. This is how the next innovation comes into existence.

While most people are satisfied with what exists- and in fact they are right to be satisfied, because what they have at hand is working for them- the golden eagle entrepreneur questions methods, processes, products, services, paradigms, and takes risks to make change and create the next big innovation.

CHAPTER SIX

RISK APPETITE

"Run from what's comfortable. Forget safety.
Live where you fear to live." **- Rumi**

Entrepreneurship is not just about spotting opportunities or being innovative; it's about having the courage to take risks. The same as the golden eagle, that dares to hunt prey much larger and seemingly out of reach, the golden eagle entrepreneur possesses a strong risk appetite.

This chapter will address how taking bold risks can lead to huge breakthroughs. We'll explore how embracing risks and stepping outside your comfort zone can spark innovation and take your business to the next level.

Risk in Entrepreneurship

In today's common language, risk generally refers to a possible negative event, i.e., a danger. This definition of risk is very narrow, as opportunities – that is, positive events – are excluded.

However, in some areas of daily life, a broader concept of risk has established itself, encompassing both negative and positive events. For instance, people talk about a 'riskier investment strategy', when comparing stocks with government bonds or a risky style of play when a soccer team adopts an offensive playing system.

In economics, the broader concept of risk is typically used, which is defined as *the deviation (both positive and negative) of a future event from the expected outcome of that event.* The term risk is used to denote both the cause and the effect of an event.

Therefore, the concept of risk can relate to any stage of a cause-and-effect relationship. For example, consider a strike that leads to a loss of sales, which in turn causes a reduction in profit.

In this case, one would speak of the 'risk of a strike,' where the strike is the immediate cause of the sales decline and indirectly the cause of the profit reduction. Similarly, it is possible to speak of a 'risk of sales loss' or a 'profit risk,' with the term 'profit risk' referring exclusively to the effect of the event.

Entrepreneurship, innovation and risk are inseparable terms that are closely tied to each other. No entrepreneurial activity can be made without innovation and risk.

Starting something new, when you know there are plenty of large companies in the industry who do things in an old fashion style, or bringing disruptions and paradigm shifts when people are used to solving their problems the old way, requires courage to take risks.

All startups have risks in their innovative practices and the risk is tied to the degree of their innovativeness. The more innovative you are, the more risk you have to take.

Risk and Uncertainty

In everyday language, the terms risk and uncertainty are often used synonymously, but strictly speaking, they represent different situations. In decision theory, there is a clear distinction between the two terms.

Risk refers to situations where we know the possible outcomes and their probabilities. For example, in roulette, the ball can land on any number from 0 to 36, with each number having a probability of 1/37. This is a risk situation because the probabilities are clear and measurable.

On the other hand, uncertainty refers to situations where we know the possible outcomes, but cannot determine their probabilities. For instance, we can say that extraterrestrial life either exists or does not, but we cannot measure the probability of either outcome.

In reality, the line between risk and uncertainty is often blurred. Decisions usually cannot rely solely on objective probabilities. Instead, they are based on a mix of subjective understanding and possible outcomes, creating a gray area between risk and uncertainty.

Risk-Readiness

In numerous contributions, risk orientation and decision-making in uncertain situations are considered constitutive characteristics for entrepreneurs.

However, this traditional view is often more of an intuitive derivation or an ideal notion of the concept of an entrepreneur, where risk-taking is often concentrated in the founding phase of the enterprise, during which the concrete prospects of success are still unclear or low.

Regarding the substantive scope of the term, almost always a specific form of risk, monetary risk, is emphasized. This view appears too narrow. Therefore, we agree with Liles, who points out that the entrepreneur takes on a variety of different risks: financial risk, career risk, family and social risk, psychological risk, and even health risk.

Risk orientation was already highlighted by Cantillon as a central entrepreneurial characteristic and later refined as the most important distinguishing feature between entrepreneurs and managers.

This assumption was confirmed in the study by Colton and Udell. In relation to startups, risk orientation, along with creativity, proved to be a better indicator than performance orientation. In the study by Jennings, Cox, and Cooper, 47% of the 21 entrepreneurs surveyed rated their risk orientation on a scale of 1 to 100 with an average value of 75, while only 32% of the 18 employed business leaders rated themselves at 50 or higher. Hull, Bosley, and Udell, as well as Atkinson, reached similar results. Atkinson found a significant difference in risk propensity between spin-off founders and R&D personnel.

McClelland discussed risk-taking in a close context with the achievement motive. In various experimental studies, he found that high achievement motivation typically combines with moderate risk-taking.

From this, it can be inferred that entrepreneurial activity is likely to involve moderate risk propensity. This assumption is essentially confirmed by Brockhaus and Sexton/Bowman-Upton. Neither study found a difference in risk orientation between entrepreneurs and non-entrepreneurs.

Delmar suggests a possible reason for this is that entrepreneurs only take risks in business decisions and, therefore, their risk-taking only occurs in areas where the entrepreneur considers themselves an expert.

The contradictory empirical results regarding the risk behavior of entrepreneurs are partly explained by the fact that the assumption of specific risks depends on the concrete situation (e.g., the life phase of the company).

Established entrepreneurs, in particular, are suspected of tending towards risk aversion or assuming limited and calculable risks due to potentially higher losses, social caution, and responsibility towards employees. Thus, risk-taking appears to be too situationally conditioned to be considered an enduring personality trait of entrepreneurs.

When considering risk-taking in the context of internal entrepreneurship, its significance clearly takes a back seat to other guiding motives, such as achievement orientation or assertiveness.

The risks that the internal entrepreneur takes are largely personal risks, such as social, psychological, and health risks. These are mainly due to their achievement orientation, which can lead them to harmful psychological and physical work efforts and to the brink of health collapse, especially in the case of failure or strong resistance against the realization of their goals. Therefore, psychological and constitutional strength are necessary traits of internal entrepreneur

Types of Business Risks

The golden eagle is known for its capacity to survey its environment with great precision and agility, the golden eagle entrepreneur should also be on his guard and act very strategically in identifying and managing the various types of risks.

Risks can be classified in multiple ways. In the business context, it is relevant to categorize risks into three main types: strategic risk, operational risk, and financial risk. Strategic and operational risks can further be divided into external and internal risks.

Strategic Risks:

External Strategic Risks

- These relate to the company's activities, markets, customer groups, and products.
- They cover all political, societal, legal, or economic developments that can impact the company.
- They apply to both current and potential fields of activity.
- This broad category includes both adverse risks and opportunities.

Internal Strategic Risks

- These pertain to the company's value chain design and decisions about outsourcing and offshoring.
- They involve decisions on resource use, such as personnel policies, procurement guidelines, and investment policies.
- Risks include quality problems, delivery failures, or theft of know-how, as well as the value generation of the company's contributions.
- These risks are general and encompass both dangers and opportunities related to competitive advantages.

Operational Risks

External Operational Risks

- These include one-off or unsystematic events that impact the company from outside.

Examples are natural disasters, legal proceedings, or fraud. Although broad and comprising both negative and positive elements, the focus is mainly on negative aspects.

Internal Operational Risks

- These involve one-off or unsystematic events initiated by internal sources.
  Examples include embezzlement, human error, IT issues, and workplace accidents.
- These are defined as narrow risks since they primarily involve negative deviations from planned processes.

Financial Risks

Market Risks

- These result from changes in market prices, such as interest rates, exchange rates, commodity prices, or stock prices.
- They can affect the company's asset positions, costs, or revenues. Market risks are general and include both adverse risks and opportunities.

Credit Risks

- These arise from the contractual obligations of partners. If a partner fails to meet obligations, it can damage the company.
- Credit risks are specific risks related to partner reliability.

Creditworthiness Risks

- These occur when a company's creditworthiness worsens, leading to potential liquidity or refinancing problems.
- They can also cause partners to refuse long-term contracts or demand disadvantageous terms, such as "delivery only after payment."
- Creditworthiness risks are narrow risks but improving creditworthiness can bring better financing conditions or contract terms.
- Despite the potential for positive aspects, creditworthiness risks are mainly concerned with negative impacts.

The mentioned risks are the business risks that entrepreneurs will face in their business sooner or later and being aware of those future risks is crucial for resilience.

Many people use the term "calculated risk", which might be strange for bold entrepreneurs who build the future and do not calculate or over analyze the risks. In the author's opinion, the risk should not be calculated, but a high degree of awareness of bold risk taking is required for entrepreneurs.

Being aware of the risk you are taking makes you mentally ready for future problems and can help you better manage future dilemmas. As the author puts it:

"The only risk calculator in the beginning of a startup journey is just being aware that you are taking risk because of your vision, and that's

necessary and enough. Too much calculation of risk kills the spirit of entrepreneurs."

While these business-related risks are pivotal to understand and manage, the entrepreneurial journey also comes with a set of personal risks that are equally significant. These personal risks can have profound impacts on an entrepreneur's life, well-being, and overall success.

Entrepreneur's Risks

Entrepreneurs take more personal risks than business risks in the beginning, but once the startup grows and becomes a structured company, there are more business risks coming.

Personal risks involve the potential impact on the entrepreneur's personal life and well-being. These risks are particularly significant because they can affect not only the individual entrepreneur but also their relationships, health, and overall quality of life.

Personal Risks

entrepreneurship involves risks that are directly imposed on the founder and they should be aware of those risks they are taking, once they decide to take the entrepreneurial action.

Stress and Burnout

Running a business is generally associated with long hours, high pressure, and continuous crises and challenges with critical decisions at every point.

This results in an amount of chronic stress, leading to possible burnout. Chronic stress can be considered a negative influence on a person's physical health, resulting in hypertension, heart disease, and a weakened immune system.

Mentally, it can affect a person's life through stress, depression, or emotional breakdowns. Stress management for entrepreneurs could involve physical exercises, meditation and relaxation, proper rest, and seeking professional help in cases where stress causes mental health problems.

Effective time management through delegation of duties and realistic goal setting can further ease the tension.

Work-Life Imbalance

The undue attention that must be given to building and sustaining a business might often result in an imbalance between work and personal life.

Entrepreneurs can overcommit time to a company at the expense of personal relationships or leisure activities. This imbalance could result in strained family relationships, low social interactions, and feelings of loneliness. It may also shrink overall life satisfaction and happiness.

Entrepreneurs need to keep work and personal time separate. Scheduling regular days off, ensuring family and friends come first, and having interests other than business will help create a better balance. Time management techniques and learning to say no to overcommitment are also critical.

Health Issues

The risks associated with overwork are significant. Chronic sleep deprivation can lead to severe health issues, reduced cognitive function, impaired decision-making abilities, and a lack of creativity.

These outcomes affect personal well-being and the success and sustainability of the business. Companies like Google and Apple recognize these risks, promoting work-life balance and encouraging employees to prioritize health and well-being to sustain long-term productivity and innovation.

Consider an entrepreneur who sacrifices sleep to put in long hours, believing this is the key to success. Over time, this lack of sleep can lead to serious health problems such as insomnia, chronic fatigue, cardiovascular disease, weakened immune system, and mental health issues like anxiety and depression.

Elon Musk has openly discussed his struggles with sleep deprivation, acknowledging that working 120-hour weeks took a toll on his health and well-being.

Insights from "Why We Sleep"

Matthew Walker's book "Why We Sleep" provides essential insights into the significance of sleep for overall health and cognitive performance.

Walker highlights that sleep is vital for memory consolidation, emotional regulation, and creative problem-solving—all critical for entrepreneurial success. The book stresses that getting enough sleep boosts productivity and decision-making capabilities, challenging the notion that sleep is unimportant for busy entrepreneurs.

By having enough sleep, entrepreneurs can maintain their health, improve their cognitive abilities, and sustain their energy levels, leading to better performance and more sustainable success. By incorporating the insights from Why We Sleep into their routines, entrepreneurs can achieve a balanced approach that promotes both personal well-being and business growth.

The myth of the overworked entrepreneur is damaging and counterproductive. Entrepreneurs must understand the importance of work-life balance and prioritize their health to achieve long-term success. A balanced approach that includes adequate sleep and self-care, as discussed in Why We Sleep, can lead to more sustainable productivity, creativity, and overall business success. By debunking the myth of endless work and realizing the value of rest and balance, entrepreneurs can truly soar to new heights.

The physical and mental demands of entrepreneurship can take a toll on an individual's health. Poor diet, lack of sleep, and neglecting physical activity are common issues faced by busy entrepreneurs.

Over time, these habits can lead to serious health problems such as obesity, cardiovascular diseases, and metabolic disorders. Mental health issues like anxiety and depression can also arise from prolonged neglect of personal well-being. Entrepreneurs should prioritize their health by adopting a balanced diet, regular exercise routine, and ensuring adequate sleep.

Regular medical check-ups and being mindful of one's mental health are crucial. Building a support network, including family, friends, and healthcare professionals, can provide essential support

Personal Financial Risks

Entrepreneurs often invest significant personal savings and assets into their business. This commitment can place their personal financial stability at risk, particularly if the business encounters financial difficulties.

This is the most common cause of stress in entrepreneurs who are fully dedicating their time to their own startup and have no other sources of income.

Financial instability can lead to stress and anxiety, affect the entrepreneur's ability to meet personal financial obligations, and potentially result in the loss of personal assets. It's important to separate personal and business finances as much as possible.

Entrepreneurs can create a financial buffer by maintaining an emergency fund, avoid over-leveraging personal assets, and seek diverse funding sources to spread the risk. Professional financial advice can help manage personal finances more effectively.

Social Risks

While personal risks affect the individual, social risks extend to the entrepreneur's network and community. Understanding and managing social risks is equally important, as these can influence both personal and business success. Entrepreneurs also face social risk during their startup journey.

Social risks pertain to the potential impact on an entrepreneur's social relationships, reputation, and standing within the community and industry.

These risks can affect how the entrepreneur is perceived and supported by others, which in turn can influence their business outcomes.

Types of Social Risks

Reputation Damage: An entrepreneur's reputation can be impacted by business failures, ethical missteps, or negative public relations.

Damage to reputation can lead to loss of trust among customers, investors, and partners, making it difficult to secure business deals and funding.

Entrepreneurs should maintain transparency, ethical practices, and strong communication with stakeholders in order to mitigate such risk. Founders need to proactively manage public relations and address any issues promptly and openly.

Relationship Strain: The demands of running a business can strain personal and professional relationships. Strained relationships with family, friends, or business partners can lead to a lack of support and increased stress. Prioritize communication and spend quality time with loved ones. Entrepreneurs should cultivate strong professional relationships through networking and collaboration.

Community Perception: How the local community views the entrepreneur and their business can affect support and success. Negative community perception can result in a lack of local support, boycotts, or negative word-of-mouth.

Founders can engage with the community, participate in local events, and contribute to community development. Build a positive local presence and foster goodwill to mitigate such risks.

Industry Relationships: The entrepreneur's relationships within their industry, including with competitors, suppliers, and regulators, are vitall.Poor industry relationships can lead to reduced cooperation, difficulties in sourcing materials, and regulatory challenges.

Founders must build and maintain strong industry relationships through networking, partnerships, and adherence to industry standards and regulations.

The Entrepreneurial Mindset for Risk-Taking

Having explored the various types of risks entrepreneurs face, it is evident that recognizing and understanding these risks is only the first step.

The true hallmark of successful entrepreneurs lies in their mindset and readiness to navigate these uncertainties by taking risk.

Just as the golden eagle doesn't merely spot its prey but also employs precision and strategic action to capture it, entrepreneurs must cultivate a mindset that embraces risk-taking with confidence and preparedness.

Confidence and Boldness

Risk-taking requires a significant degree of confidence and boldness. Entrepreneurs must believe in their vision and have the courage to pursue it, even in the face of uncertainty.

This boldness drives them to innovate, explore new markets, and challenge the status quo, ultimately leading to growth and competitive advantage.

Risk-Taking vs. Thrill Seeking

Entrepreneur's passion and vision are the motives behind risk-taking. Future foresight and building the future make the entrepreneur take risks.

However, risk-taking is not the same as thrill-seeking which is impulsive and based on short term gambling without considering consequences.

The golden eagle never attacks a human in the city. Entrepreneurs should also be aware not to take impulsive thrilling decisions that puts them and their business at extreme danger. For example, doing illegal acts with prior knowledge is not risk taking, it's thrill seeking. Here is another example by the author:

"I have seen many thrill seekers and gamblers in the business world. One of the most common thrill seeking startups do is when they have

problems, in particular financial problems, they step into unethical and even illegal acts to solve the current problem and they bring a chain of future problems into their company by misconduct.

In one case a startup stole a client list of a competitor, then the competitor which was new and in less power realized the hacking and then asked the large company to give them their full list of clients in return for not taking any legal action or publicly destroying their unethical business conduct.

Once we were doing due diligence of the startup, we heard about this from one of the ex-employees of them and we dismissed the investment. Thrill seeking will have long-term problems and you should know that people eventually talk."

The golden eagle entrepreneur has a clear vision and the passion to reach that vision is so high that taking risks is accepted as an inevitable part of his startup journey.

The golden eagle entrepreneur tries to be prepared for the tough times, by first having a high degree of awareness that a startup activity requires and involves risk. Equipped with this awareness, he follows the passion and during the journey stays away from thrill seeking and gambling for short term results.

CHAPTER SEVEN

ENTREPRENEURIAL FOCUS

"Let yourself be silently drawn by the strange pull of what you really love. It will not lead you astray." - Rumi

Once the golden eagle identifies an opportunity, it locks onto its prey and begins the hunt. Similarly, the golden eagle entrepreneurs epitomize the power of focus in the business world.

With a sharp eye for opportunities, they soar to great heights and zeroes in on core strengths and strategic goals. They prioritize solving key problems and upholding core values, rejecting distractions. Let's discuss entrepreneurial focus as we proceed in this chapter.

The Power of Focus

When the author was studying in Sweden, he faced a surprising change in the education system. Here is the story:

"The Swedish education system was the first shock that I faced during my studies in Umeå university back in 2009. We had to study one course every day for a month and after finishing that single module and passing the exam, we started the next one.

This was strange for me, as I came from an education system in which students dealt with multiple courses and topics and instructors every day and took exams one after the other at the end of a semester.

At that time, I thought they made a mistake, because people should learn how to deal with different tasks everyday when they are in professional work. But I also considered that they also have world class successful companies that are all made with people who did their whole life studies in this education system.

Companies such as Ikea, Volvo, Skype, Spotify, Klarna, etc. all started by people who studied, lived, and worked the same way. This interesting topic made me dig into this system and I found that the key point is focus. The Swedish education system simply asked us to focus on one task at a time.

One task at a time, one idea at a time, and one market at a time is simply what their system asked us. In the end we used the collective know-how gained by this system and used them accordingly."

Entrepreneurial focus begins with a clear vision. As mentioned earlier, the golden eagle entrepreneur has a vivid picture of what they want to achieve and the impact that they would want their business to have.

This will be the guiding star; it's directly leading every effort and decision to one goal. A clear vision will make these entrepreneurs remain motivated even in the midst of challenges and failures. This gives them a sense of purpose that is in line with their long-term objectives.

Another important aspect of entrepreneurial focus is setting SMART (*specific, measurable, achievable, relevant, and time-bound*) goals. Dividing the bigger picture into smaller, more manageable building blocks enables founders to define a road map for success.

SMART goals are very clarifying and structured in such a way that a founder can really monitor if he/she is on course or if an adjustment needs to be made. This targeted approach assures that all actions made are meaningful and directly lead to business growth.

Discipline underpins focus in entrepreneurship. The entrepreneurship path is filled with distractions, from new opportunities that may not add value to the core of the business and form the future of the company.

Founders who are disciplined and focused are able to stick to their plan of doing things that will take them to their goal and push off or put aside activities that distract them from achieving their goals.

It enables founders to reap most of their productivity and ensure all efforts are focused on what really matters. As the author experienced, most founders, being excellent at identifying and exploiting opportunities, lose focus when new opportunities come during their entrepreneurial journey. Here is how the author puts the experience with these founders.

"I saw most founders get excited with new opportunities, which in most cases, I can say, could be just opportunity traps. One of our AI startups got excited when a prospect asked them to develop features for them.

The startup business model was SaaS and they already had the MVP up and running and was good enough to hook some customers. The demand from the client was too much, as what the customer wanted was full customization.

I asked them to negotiate and start with what the MVP is with some minimum customizations because that was a pitfall for their growth.

SaaS startups, in particular at the early stage of their firm, have a high potential to become consultants, advisors, or just freelancers if they don't follow product orientation and customize too much or care too much for one client.

This is one of the hardest tasks for bootstrapping entrepreneurs as they are seeking extra income to put in their company, but at the same time they should say no to some clients and stay focused."

Saying no is the ability that determines startup success, and every founder must master it. The golden eagle entrepreneur knows that it is not possible to chase two rabbits at the same time.

More often than not, they must see and analyze which opportunity would be best served in the light of strategic alignment with their vision and goals.

By saying no to distractions and noises, they are able to focus their resources and energies on what is critical to them: innovating and growing their innovation. This selective ability would keep the business focused and avoid distractions.

Time management is another key entrepreneurial concern area. Many times, entrepreneurs have to multitask and be good at several skills.

On that end, they must be keen to learn how to focus, manage time through task prioritization and set appropriate schedules, and avoid procrastination.

Through techniques, for example, time blocking, setting task deadlines, and minimizing interruptions, entrepreneurs can maintain sanity during planning while allocating the time available to activities that make the most impact. Effective time management guarantees sustained concentration and efficiency.

On the aspect of entrepreneurial focus, this approach must find its power from the support system. Having common vision in the team is also another aspect that helps staying focused.

When all members of a team are aligned and working towards common objectives, it magnifies the effort of an entrepreneur and enhances the business's chance to get nearer towards its goals.

Essential elements that will help achieve such a task include effective communication, strategic goal-setting sessions, and creating a collaborative culture, which keeps all focused and motivated.

Focus vs. Obsession

Focus is the ability to develop and sustain a clear vision, continuous attention, and mental as well as physical effort toward an intended goal or objective.

For a founder, that means having a clear vision of what you're trying to achieve with your business, keeping a balanced perspective, staying flexible to feedback and changes in the market, and exercising control over your actions without being consumed by them.

Focused entrepreneurs strive to set clear objectives, maintain a healthy work-life balance, flex their strategies as appropriate, and make conscious efforts at concentrating on tasks.

Obsession, on the other hand, is an overwhelming preoccupation with a particular idea, task, or goal that tends to dominate thought and behavior.

Obsession in the context of entrepreneurs is an intensive, almost ungovernable concentration on the objectives and goals in business, usually characterized by the omission or subtraction of other important needs such as personal relationships, health, and leisure.

Obsessive founders show a high level of dedication, an inability to adapt or let go even when it is necessary, and an urge to constantly think about and engage in business activities which are often driven by anxiety or fear of failure that could turn out to be fatal in the long run, for both the founder and his start-up.

There are several positive aspects of focus for entrepreneurs. It boosts productivity, the possibilities for achievement of the business objectives on time, and maintains an equal sense, hence leading to satisfaction.

Focused entrepreneurs are prone to innovation of new solutions, established strong connections, and sustainable business practices.

On the other hand, fixation can accomplish more because of high incentive, focus, and a firm will to overcome barriers. Obsessive entrepreneurs could achieve unbelievable feats and deliver quality results due to their perfectionist drive.

Too much focus can also lead to negative results. Founders who are too narrowly focused might miss out on new opportunities or ideas, risk burnout if they maintain intense concentration for too long without breaks, and experience increased stress levels.

Obsession's negative aspects are more severe, including a high risk of physical and mental exhaustion due to relentless pursuit of business goals, neglect of personal relationships and health, inability to adapt or change course, leading to potential failure or missed opportunities, and increased anxiety and stress due to constant preoccupation with business success.

Practical examples illustrate the differences between focus and obsession in the entrepreneurial world. A focused founder sets clear priorities for the business day, dedicates specific time blocks to tasks, takes regular breaks, and maintains a healthy work-life balance.

Conversely, an obsessed entrepreneur works excessively long hours, neglecting personal health and relationships, driven by an overwhelming need to make their business succeed.

A focused entrepreneur creates a balanced schedule that includes dedicated work sessions, breaks, and time for personal activities, ensuring sustained productivity and innovation. An obsessed entrepreneur spends every waking hour working on their business, unable to relax or engage in social activities, driven by the fear of failure.

In the following table, we demonstrate the aspects of focus and obsession in entrepreneurship.

Aspect	Focus in Entrepreneurs	Obsession in Entrepreneurs
Definition	Concentration on business goals with clarity and sustained effort	Overwhelming preoccupation with business goals
Characteristics	Clarity, balance, flexibility, control	Intensity, imbalance, rigidity, compulsion
Positive Aspects	Productivity, achievement, well-being	Drive, detail-oriented, resilience
Negative Aspects	Limited perspective, overwork, stress	Burnout, neglect, inflexibility, anxiety

Entrepreneur Scenario	Sets priorities, balances work and breaks	Works excessively, neglects personal health and relationships
Startup Scenario	Balanced work schedule with breaks and personal activities	Constant work, neglects relaxation and social activities
Balancing Strategies	Set boundaries, practice self-care, seek feedback	Often lacks balance, driven by anxiety and fear

There's a myth in the entrepreneurial world that success requires excessive work hours, often at the expense of personal health and relationships.

While dedication and hard work are important, the notion that endless work is the only path to success is both misguided and harmful.

Entrepreneurs who adhere to this myth risk burnout, decreased productivity, and deteriorating health, ultimately jeopardizing their business.

The Optimum Choice for the Golden Eagle Entrepreneur

To achieve a healthy balance, it is essential for entrepreneurs to recognize when focus is tipping into obsession and take steps to manage it.

This can be achieved by setting boundaries, clearly defining work and personal time to ensure a healthy balance, practicing self-care, incorporating regular exercise, relaxation, and hobbies into their routine, seeking feedback, regularly seeking input from mentors and peers to maintain perspective and adjust focus as needed, staying flexible, being open to changing goals and approaches based on new information and circumstances, and monitoring well-being, regularly assessing physical and mental health to identify signs of burnout or stress.

Focus enables golden eagle entrepreneurs to enhance productivity, achieve business goals, and foster well-being without the detrimental effects of obsession. It allows for innovative solutions, strong networks, and sustainable practices, all crucial for long-term success.

Obsession, while potentially driving short-term achievements through sheer intensity and attention to detail, poses significant risks such as burnout, neglect of personal relationships, and an inability to adapt.

These factors can ultimately undermine both personal health and business success. The golden eagle entrepreneur requires to maintain a balanced approach in their entrepreneurial focus, avoiding long-term obsessions.

Competitor Focus vs. Customer Focus

Just imagine this, the golden eagle is hunting and while scanning the ground from above to hunt hares, it sees another bird of prey in the vicinity at the same time that it finds a prey.

What happens if it focuses on the other bird of prey and not the prey? What if both focus on each other and not the prey?

In the era of highly competitive markets, most people focus on competitors and this focus is so high that they forget one of their most important stakeholders that they should focus on: the customer.

There are two types of companies: Some focus on competitors and want to follow them or get better results and some focus on customers and deliver the most value to their customers. Amazon has been known for being customer centric that later they called it customer obsession. Amazon thought and invented on behalf of customers and they brought many new innovations with that approach.

The golden eagle always focuses on the prey and so does the golden eagle entrepreneur by focusing on the customers and what they can deliver to them. The golden eagle entrepreneurs never focus too much on competitors, instead, they invest and innovate to meet all customers needs.

Growth, Innovation, and Entrepreneurial Focus

When it comes to entrepreneurial focus, there are some aspects of startup activities we should consider. There are majorly two parts that are crucial to consider when it comes to focus. A founder might ask these questions: If we have to focus then *how can we innovate? How can we scale?*

Innovation is the heart of entrepreneurship and startups should consider innovating. It is essential to keep focus and innovate and grow at the same time.

Focus is not "not innovating" and "not growing", but focus is the fuel for both innovation and growth. Many people mistake distraction with growth and innovation, but innovation and growth especially at the early stages of your startup should happen in your core value and core of your innovation. In other words, you have to focus on your core innovative product and try to do your first scale in that area.

This is very critical especially in the pre-seed and seed stage of startups to stay focused. At this stage, distractions come and opportunity traps are all around you. Let's hear another experience from the author:

"I always faced an issue of people not being able to deliver an elevator pitch. The core of this problem is not because they are not smart or shy; it is because they did not find their core value themselves yet.

An elevator pitch is the best tool that can help you find your own focus. If you can not pitch the core of your idea or innovation in 60 seconds, you have not found it yet.

During execution, I also faced that startups are hurrying to grow or innovate, when they are not ready for that. They often spend money to do a nose job when they need a heart surgery."

Entrepreneurial focus is not only for startups. Larger corporations are also facing these paradoxes of innovation and growth and the answer always lies in the core value.

A corporate that wants to implement an open innovation practice should check how far from the core activities they want to go and whether the goal is strategic or financial. Or a corporation that wants to enter a new market, should consider how close to the core innovation or activity is the new market and what the objectives and strategies they have by entering the new market.

Focus of the Golden Eagle Entrepreneur

The majestic golden eagle uses a high degree of focus on each prey during the hunt, but when it fails, after accepting the failed hunting, focuses on the next hunt. The golden eagle does not obsess on the prey when the prey is totally out of reach and does not starve because of this obsession.

The golden eagle entrepreneur also focuses on a fundamental trait that drives business success. The first focus of the golden eagle entrepreneur is the customers and your delivering the most value to them.

By having a clear vision, setting SMART goals, saying no to distractions and noises, believing and staying on the core value he will soar to the heights of innovation and entrepreneurship.

By cultivating these aspects of focus, the golden eagle entrepreneur can navigate the complexities of the business world, achieve objectives, and create lasting impact. The golden eagle entrepreneur also maintains an optimum choice in his focus to make sure it does not lead to a long-term obsession.

CHAPTER EIGHT

BOLD INITIATION & DECISIVENESS

"As you start to walk on the way,
the way appears." — **Rumi**

Golden eagle entrepreneurs are fearless in taking the very first step, following the footsteps of the golden eagle, known for its daring, fearless, decisive, and bold dives and actions. They possess extraordinary abilities to launch ventures with confidence and clarity, navigating the uncertainties with a firm hand.

By boldly initiating, they set the stage for innovation and growth, turning ambitious visions into reality. This chapter explores the fearless beginnings of golden eagle entrepreneurs, highlighting the bold actions and decisive strategies that drive their success.

On Being Bold

Boldness is the quality of being courageous, confident, and unafraid to take risks. It encompasses a fearless approach to challenges, a willingness to push boundaries, and the readiness to act decisively in uncertain situations.

Bold individuals do not shy away from making tough decisions and are prepared to step outside their comfort zones to achieve their goals.

This trait is characterized by a proactive attitude, a strong belief in one's vision, and the resilience to persist through setbacks and failures. Courageous decisiveness is the key element of being bold and the golden eagle entrepreneur makes fast and confident decisions.

There are three types of people in the world when it comes to startup initiation. Those who don't recognize the opportunity and obviously don't take any action, those who recognize the opportunity and don't take any action, and finally those who recognize the opportunity and take action.

No matter whether you explored the opportunity or not, there is nothing entrepreneurial, if you don't take any action. In other words, those who explore the opportunity and didn't take action are not different from those who didn't even recognize the entrepreneurial opportunity.

Type 1	Type 2	Type 3
Explores the opportunity	Doesn't explore the opportunity	Explores the opportunity
Doesn't take any action about it	Doesn't take any action about it	**Takes risks and acts**

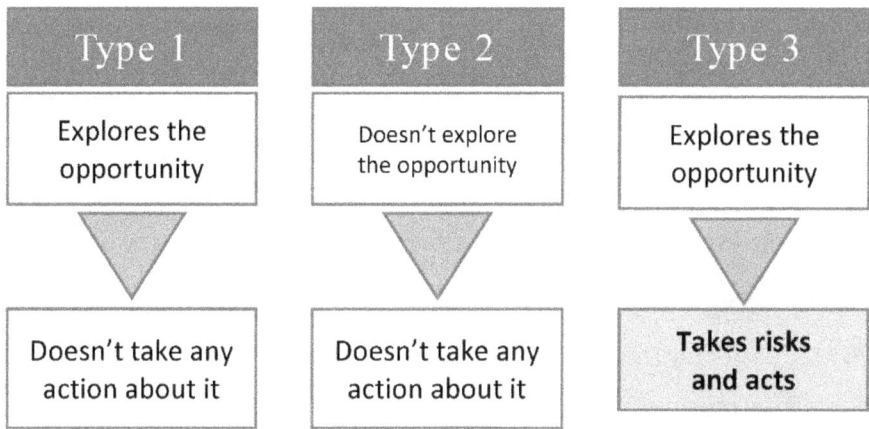

From these three types it is obvious that we should not discuss type 1 and 2 here.

We all have seen many people of type 1 and there are a variety of reasons why they didn't take any action about the explored opportunity.

In some cases they don't think it's wise to talk about their business idea and kept the SECRET idea with themselves, which almost in most cases only stayed in their mind forever. Here's the experience of the author confronting them:

"I sometimes offer limited free office hours for those who need advice on their startup or business idea and in those sessions I faced this problem.

There are many people who think the idea matters, and I found out that those who are tightfisted about revealing their business idea hardly become entrepreneurs.

It's not only a matter of trust, but it is also about networking, team building, finding resources and knowing people who will be helpful on the way. In those office hours sessions, there were people who were generous enough to openly talk about their business ideas and as their ideas were not fitted for our investment domains, I introduced them to people who could help them or be their partner. Some people think a lot about the essence of the idea and overlook the action and when I face them, I usually answer to these tight people that 'I also have an idea of flying to Mars this afternoon and have my dinner there and come back tomorrow morning for the meeting I have, but can I really do that?" And this way I try to make them understand the importance of DOING something now rather than keeping that idea with themselves.'"

An entrepreneur's leap of faith is a vital element of entrepreneurship. Those who pursue their dreams have a deep belief in something that seems absurd to most people.

This leap of faith is something that many people don't understand. While others see reasons, current market conditions, and risks, the entrepreneur sees the future and decides with heart and guts.

With the help of entrepreneurial subjectivity, they become risk-blind positive thinkers who lock on to their target, and are unstoppable in their pursuit. This leap of faith is backed by entrepreneurial bias long before any VC or angel investor knows about the idea. This bias is tremendously active during the opportunity recognition phase, driving action that has led to many innovations in human history.

Leap of faith is something that not only entrepreneurs but also the VCs backing them should possess. Entrepreneurship speaks the same language and has the same nature worldwide. No matter where you practice entrepreneurship, it remains the same everywhere.

An entrepreneur's leap of faith is built on the foundation of opportunity recognition. It is the bias and entrepreneur's intuition that leads to recognition, and those who recognize unique, unprecedented opportunities are willing to take action.

Beyond Faith

Passion is beyond just faith and believing in the business idea. It takes you to the next level, which is literally taking action.

With passion, you have the highest degree of focus and that's what can only be fulfilled by taking risks and initiating. Here are experiences of the author:

"When I started Lar Fishing in Sweden in 2010, I put all my money in that online store. I listened to people who told me about the reasons why at the crisis time,a vertical online shop that sells niche products for fly fishers is a terrible idea, but I was so biased that I never took heed to any of them. I remember when I went for a market study in Stockholm fly fishing fair to see what percentage of people buy these products online, I came with a result that showed more than 80% of fly fishers in Sweden want to feel the product and try them first and are not willing to buy online.

If I was going to act logically based on facts and figures, Lar Fishing should have never started. I didn't listen to any of those people because

I was thrilled with the business idea and the fact that I should start it ASAP.

Then I started the web site and did everything I could do to have everything ready. I used to study during the day and work on Lar Fishing after school. I lost sight of the clock and worked hard to get what I wanted to see. That was the passion I cannot really describe here.

The same passion happened to me when I co-founded a corporate innovation center in 2018, and on top of all, when I started in 2023 by investing all my years of savings into it. When you do something with your heart and passionately, you don't calculate much and you will never regret."

Entrepreneurship is often lauded as a journey of innovation, risk-taking, and resilience. However, the path to success is fraught with mental barriers that can block progress. There are several traits that block the journey of entrepreneurship and paralyze the entrepreneurs.

Paralyzers of Entrepreneur

There are a variety of traits that paralyzes entrepreneurs from taking action or even making a decision. Knowing about the blockers of entrepreneurial initiation can help you get a better understanding of the root causes of not acting.

As experienced by the author for many years, these are also problems to the larger companies and their executives as well. Let's dig into the paralyzers of entrepreneurship.

Perfectionism

Perfectionism refers to the tendency of entrepreneurs to set exceptionally high standards for themselves and their ventures, driven by a desire for excellence and success.

While this trait might fuel innovation and growth, it also poses significant challenges and risks to both the entrepreneur and their business.

On the positive side, perfectionist entrepreneurs often produce high-quality products and services. Their meticulous attention to detail ensures that offerings are well-developed, reducing the likelihood of errors and increasing customer satisfaction.

Additionally, driven by a constant desire to improve, these entrepreneurs are never complacent. They seek out feedback, identify areas for enhancement, and strive to refine their operations and products, fostering a culture of excellence and innovation.

This relentless pursuit of perfection often translates into a strong work ethic. Perfectionist entrepreneurs are highly dedicated and committed, inspiring their teams and leading to a highly motivated and productive workforce.

By consistently delivering top-notch products and services, they can build a strong reputation for quality and reliability, which can be a significant competitive advantage, attracting loyal customers and partners.

However, entrepreneurial perfectionism also brings several challenges. The quest for perfection can lead to procrastination, with entrepreneurs delaying product launches or decisions until everything is "just right". This can result in missed opportunities and slow market entry. Let's hear the author's experience of the perfectionism in startups:

"Perfectionism, in my opinion, is the killer of startup and entrepreneurial spirit that in higher degree acts as one of the main reasons for startup failure. I had a friend who was a graphic designer and one of the best designers and art directors I have even seen.

He had an idea for a startup, talked to me and I agreed to help them for a small equity and put some hours per week on their startup. We started the work and the development started. We used to meet weekly and talk and talk, but things went on very slowly. We worked for more than 9 months and always waited for a PERFECT product to be launched.

He was never satisfied and always wanted a better and better version. After a long time, the startup was shut down before launch! Yes! We failed before even launching our product just because of perfectionism. I have seen many potential entrepreneurs who wait for the perfect timing, perfect product, perfect team, perfect financial conditions, and other perfects to launch their products.

Not only this causes a delay and procrastination in launching the product, but also during the startup journey, being perfectionist stops you from taking risks and makes the whole team cautious. You will do over-analysis in order to make the best decision, that I bet in 95% of cases you will never make.

As an entrepreneur or a corporate leader, you will inject this mindset into your whole company and build a perfectionist culture that kills innovation, creativity, execution, and the golden eagle spirit in your company."

The intense pressure to meet high standards can also lead to stress and burnout. Entrepreneurs may overwork themselves, neglecting their physical and mental health, which can ultimately impair their productivity and decision-making abilities.

Furthermore, perfectionist tendencies can lead to micromanagement, stifling creativity and autonomy among team members. This creates a work environment where employees feel undervalued and disengaged, potentially leading to high turnover rates.

Inflexibility is another issue associated with entrepreneurial perfectionism. A perfectionist mindset could make entrepreneurs resistant to change and experimentation. Fear of failure has a direct relationship with perfectionism.

It can also hinder your ability to adapt to new circumstances, limiting the innovation potential and responsiveness to market dynamics.

Overemphasis on perfection can also result in overanalysis and indecision. Entrepreneurs might struggle to make timely decisions, getting caught up in details rather than focusing on the broader strategic picture.

Embracing imperfection and understanding that mistakes and failures are part of the entrepreneurial journey can foster a more resilient and adaptable mindset. When you jump into the world of entrepreneurship, you should face your fears and embrace failure.

Learning from these experiences can lead to valuable insights and improvements. To manage entrepreneurial perfectionism, setting realistic goals is essential.

Entrepreneurs should aim to set achievable and realistic benchmarks for success, recognizing that perfection is often unattainable.

But in case you want to shoot for the moon, you should also send this message to your team that you are not afraid of failure. Delegating tasks and trusting team members can alleviate the burden of perfectionism. Empowering employees to take ownership and make decisions can enhance their engagement and productivity.

As the author puts it here:

"If an organization is always hitting the goals and does everything as planned or delivers as expected, there is a hidden killing bacteria inside this organization.

For me, if someone is always doing things right without making mistakes, this means that I have the worst employee. I believe entrepreneurship is not for entrepreneurs and all organizations and people who work in companies should act like entrepreneurs.

When you do things right all the time, you are definitely stuck in your comfort zone doing nothing new. If you stay perfectionist and try to do one thing meticulously all the time and do not explore new things, your life is on auto-pilot.

Every organization with this culture is destined to failure. If you avoid smaller affordable failures in your company, you will definitely face a fatal failure at a much larger scale in the future."

Imposter Syndrome

Imposter syndrome in entrepreneurship is a psychological phenomenon where entrepreneurs feel unworthy of their success, doubting their abilities and fearing that they will be exposed as frauds.

This pervasive sense of self-doubt and inadequacy can be particularly crippling in the high-stakes world of entrepreneurship, where confidence and decisiveness are crucial. Entrepreneurs experiencing imposter syndrome often attribute their achievements to luck rather than their skills or hard work, which can severely undermine their self-esteem and professional growth.

The paralysis caused by imposter syndrome manifests in various detrimental ways. Entrepreneurs may hesitate to take bold steps or pursue new opportunities, fearing they will fail or be unmasked as impostors. This hesitation can stifle innovation and slow the growth of their business.

Furthermore, the constant self-doubt can lead to over-preparation and perfectionism, where entrepreneurs spend excessive time on minor details to avoid making mistakes. This not only hampers productivity but also delays critical decision-making and project completion.

Imposter syndrome can also strain relationships within the business. Entrepreneurs struggling with this condition may find it difficult to delegate tasks, fearing that their lack of competence will be revealed.

This can lead to micromanagement and a lack of trust in their team, creating a toxic work environment that stifles creativity and collaboration.

Additionally, the emotional toll of imposter syndrome can lead to burnout, as entrepreneurs push themselves beyond their limits to prove their worth, often neglecting their mental and physical health in the process.

Addressing imposter syndrome requires a multifaceted approach. Entrepreneurs should seek support from mentors, peers, or professional networks to gain perspective and validation of their achievements.

Acknowledging and celebrating successes, no matter how small, can help build confidence and counteract feelings of inadequacy.

Additionally, focusing on continuous learning and personal development can reinforce an entrepreneur's sense of competence and reduce self-doubt. As an entrepreneur who already took action you should know that you are already among the top minority who dared to take risk and action and also you should know that you should not know everything yourself.

Building confidence is the key to overcoming this problem. This can be done by promoting yourself or taking extra learning materials. As mentioned before, entrepreneurs are addicted to learning over and over and this is the way of the golden eagle entrepreneur to keep training themselves through books, podcasts, and other learning materials.

Overwhelm and Indecision

Entrepreneurs often face a multitude of challenges, including the constant pressure to make critical decisions and manage a wide array of tasks.

This can lead to feelings of overwhelm and indecision, which can be particularly paralyzing for even the most ambitious and driven individuals. Overwhelm and indecision are common obstacles that can significantly impede an entrepreneur's ability to effectively lead their business and make progress toward their goals.

Overwhelm occurs when entrepreneurs are confronted with an excessive number of tasks, responsibilities, and decisions, leading to a state of mental and emotional overload.

This can result from trying to juggle too many aspects of the business at once, from managing operations and finances to marketing and customer relations. The sheer volume of work can create a sense of chaos and anxiety, making it difficult to prioritize and focus on what truly matters.

Indecision, on the other hand, is often a byproduct of overwhelm. When faced with too many choices or an unclear path forward, entrepreneurs may struggle to make timely and effective decisions.

This can be exacerbated by the fear of making the wrong choice, leading to analysis paralysis where decisions are delayed or avoided altogether. Indecision can stall progress, allowing competitors to gain an edge and potentially causing missed opportunities.

Entrepreneurs and corporate executives need to make fast decisions with whatever is available. This is what the author also experienced in both deal selection and closing the deal in most corporate venture capital firms and has been one the main problems in VC and startup relationships. This happens mostly when the VC decision makers are not focusing on startups only and are overwhelmed with many other tasks at the same time. This sort of indecision or late decision is also related to the focus that we covered in chapter seven.

Decision fatigue is also another term we should consider here. Decision Fatigue is the decline in decision-making quality after a long session of decision-making, leading to poorer choices as mental energy depletes.

This phenomenon can affect anyone but is particularly challenging for entrepreneurs who constantly face critical choices. The cognitive load of making numerous decisions can result in impulsive choices, procrastination, and decision avoidance.

Need for Control

The need for control is a prevalent trait among entrepreneurs, characterized by a desire to manage every aspect of their business meticulously.

While this drive can ensure high standards and attention to detail, it often hinders effective decision-making and action-taking, ultimately stifling business growth and innovation. It is also one of the main reasons according to the author's experience:

"Most people do not take any action in the beginning. Many potential entrepreneurs will not find co-founders or are not able to build a team as they want full control or think they can do it alone.

This was one of my mistakes in the past during my own startup journey that always wanted to have full control over what I am doing. In Lar Fishing in Sweden, I was a one-man army and rejected many co-founders as my partner.

After some time, I faced many challenges, I was overwhelmed, burned-out, tired, and decision fatigue also was there. By entering into the world of VC and working for years as a venture capital, I changed.

I think the best role models for delegation and being able to trust and lose control are pre-seed and Seed stage VCs, in which you do not have much control over your money and you delegate your future value creation to the entrepreneur.

Later, when I was CEO of a corporate accelerator, I built a team of 25 people and delegated the tasks much easier. Then I had time to think, to plan, to strategize, and to actually do what a CEO should do."

Entrepreneurs with a strong need for control tend to micromanage, focusing on minor details rather than delegating tasks.

You will find a close similarity when it comes to controlling and perfectionism. Controlling slows down decision-making processes, causing bottlenecks and inefficiencies.

Additionally, the reluctance to delegate stems from a fear that others might not meet their high standards, preventing them from leveraging their employees' skills and expertise, which limits creativity, innovative solutions, and diverse perspectives.

The need for control can lead to overanalysis, where excessive time is spent evaluating outcomes before making decisions, resulting in analysis paralysis. Entrepreneurs may also become risk-averse, fearing deviations from their plan, thus avoiding necessary risks essential for innovation and competitive advantage, causing their businesses to become stagnant.

The constant need to oversee all aspects of the business can reduce agility, making it harder to respond to market changes and seize new opportunities promptly. This hands-on approach can also lead to significant stress and burnout, impairing decision-making abilities and overall performance. Furthermore, it can stifle innovation within teams, as employees may feel discouraged from sharing new ideas or taking initiative due to micromanagement.

Unparalyze Yourself

To overcome these mental barriers, entrepreneurs can adopt several strategies. Embrace imperfection by accepting that mistakes and failures are part of the entrepreneurial journey.

Focus on progress rather than perfection. Cultivate a growth mindset by viewing challenges as opportunities to learn and grow. Prioritize and delegate tasks effectively, allowing for more efficient operation and growth.

Seek support from mentors, peers, or coaches who can provide guidance, support, and perspective. By recognizing and addressing these mindsets, entrepreneurs can unlock their potential, take decisive action, and navigate the complexities of their entrepreneurial endeavors with greater confidence and success.

To mitigate these negative impacts, entrepreneurs need to build trust in their team members, delegating tasks to free up their time for strategic planning and innovation. Establishing clear goals and boundaries helps reduce the need for micromanagement, fostering a sense of accountability within the team.

Adopting a growth mindset, where failures are seen as learning opportunities, encourages continuous improvement and innovation. By focusing on high-level strategic priorities rather than day-to-day operations, entrepreneurs can drive more impactful decisions and business growth.

Engaging with mentors or advisors can provide valuable insights and support, helping entrepreneurs navigate challenges, make informed decisions, and balance their need for control with the necessity for delegation and trust in their team.

While the need for control can drive high standards and operational excellence, it often hampers effective decision-making and action-taking. By learning to delegate, building trust in their team, and focusing on strategic priorities, entrepreneurs can overcome these challenges and lead their businesses more effectively.

Paralyzer	Symptoms	Impacts	How to Overcome
Perfectionism	Procrastination, micromanagement, stress, and burnout	Delays in product launches, slow market entry, stifled creativity, and high employee turnover	Embrace imperfection, set realistic goals, delegate tasks, and foster a growth mindset
Imposter Syndrome	Self-doubt, fear of failure, over-preparation, and avoidance of new opportunities	Stalled business growth, strained relationships, and burnout	Seek support, celebrate successes, focus on continuous learning, and build confidence
Overwhelm and Indecision	Mental overload, analysis paralysis, and delayed decisions	Missed opportunities, slow progress, and competitive disadvantage	Simplify choices, prioritize tasks, set time limits, and take regular breaks
Need for Control	Micromanagement, reluctance to delegate, and overanalysis	Reduced agility, innovation stifling, and high stress	Trust team members, set clear goals, adopt a growth mindset, and seek mentorship

Entrepreneurial Paralyzers and how to solve them

Courage and Boldness

The golden eagle entrepreneur is not paralyzed by the fear of the unknown future. They recognize that uncertainty is an inherent part of innovation and progress.

Instead of waiting for perfect conditions, they are willing to dive into uncertain conditions, understanding that this is where the most significant opportunities often lie. This courage to embrace uncertainty allows them to explore new markets, develop novel products, and disrupt existing industries.

Like the golden eagle that dives swiftly and decisively to capture its prey, entrepreneurs must be bold in their risk-taking and initiative. This means having the courage to make tough decisions, even when the outcomes are uncertain. While analysis and preparation are crucial, there comes a point when action is needed.

Being overly cautious can lead to missed opportunities, as the competitive business landscape often rewards those who move quickly and decisively.

Entrepreneurs who embody the spirit of the golden eagle balance careful planning with decisive action. They are brave, bold, and ready to take on the risks that come their way, ensuring they are prepared to navigate uncertainties with confidence.

These entrepreneurs understand that while preparation minimizes risk, it is the willingness to act despite uncertainty that drives true innovation and success.

CHAPTER NINE

NETWORK

"Set your life on fire. Seek those who
fan your flames" — **Rumi**

Building a strong network is key to making it big as an entrepreneur. Think of it like a golden eagle - it soars to great heights by being super observant and strategic about interactions in its environment. In the same vein, entrepreneurs can use their networks to gain valuable insights, resources, and opportunities.

In this chapter, we'll break down why networking is so important for entrepreneurs, how it can make or break your venture, and how to build and use a network like a pro (using the golden eagle as our metaphor).

Building a good network is the key in building a sustainable business and that's why one of the main tasks of startup founders is to build a network of partners and stakeholders around their company. Your network can be helpful in seeking advice, strategic partnerships, funding and investment, building your team, entering new markets, and every other single purpose you may have in your business. Although having a specific purpose facilitates the creation of meaningful connections, building a network is a long-term process.

The Entrepreneur's Network

When it comes to the entrepreneur's network, we have to consider, inside, and outside, on the top and down and left and right of the whole startup value chain. Here are the major networks that a golden eagle entrepreneur should consider.

Your Own Team

Your own team is your primary network, inner circle and your ride-or-die crew, and the most crucial part of your network. And yet, many new entrepreneurs forget to prioritize them.

Think about it, your team is the foundation of your venture. They're the ones who will help you turn your vision into a reality. So, it's essential to build a strong, supportive, and skilled team that's got your back.

Don't overlook the power of your own team! They're the ones who will help you navigate the ups and downs of entrepreneurship, provide valuable feedback, and help you grow both personally and professionally.

Invest in your team, and they'll invest in you and your venture. It's a win-win!

Your team is the heart of your business and the most important criteria during your fund-raising for the investor is the team. Investors value your team more than anything else.

Your Co-Founder

When choosing your co-founder, make sure to consider some important facts.

As experienced by the author, most founder and co-founder unsolvable problems and issues occur in the *why* of them becoming a part of the founding team.

This in particular happens when a founder with a passion and vision initiates a startup and months later adds a co-founder without digging into the personal agenda of the co-founder.

If you start with the "why" before you add the co-founder, you will have a better understanding of what could come next and if you can work well together or not.

This is the key element of starting with the agenda and their purpose. Some people come to you with a hidden agenda and it is important to find out and also have the courageous authenticity to say NO to a co-founder whose agenda is not matching yours.

If you have the same agenda with your co-founder and you have some conflicts, then it is surely solvable. In case you all start together, make sure that the "why" is the same for all of you.

Another factor that you should take into account is to build trust and if the co-founder is the person with the same mission working with you to build the future, then share as much as possible and make the co-founder be part of the team. Building trust starts with you first, and as the founder you are the one who should take responsibility and build this trust.

What was experienced is that one golden eagle entrepreneur is enough in each team. If the team has one golden eagle entrepreneur the leadership of that entrepreneur will be the fuel for the whole team.

External Co-Founders

Your internal team is the backbone of your startup, but don't sleep on the power of external co-founders!

External co-founders can bring in fresh perspectives, diverse experiences, and complementary skills that can significantly enhance your startup's potential.

The challenge, however, lies in finding and integrating the right external co-founders who align with your vision and can add substantial value to your entrepreneurial journey.

An external co-founder can take your startup to the next level with a completely new idea, experience from other industries, and networks that will open up doors for opportunities that might never be taken seriously. Onboarding cofounders is done after careful thought and strategic integration. Most of the time, an external mind brings in specific skills that fill the gaps in a current team's capability.

It might be technical ability, it may be marketing acumen, or it might be strategic thinking: it is these contributions that can potentially go a long way in giving your startup the added heft that it needs.

For example, if your strength is in tech and product development, an external co-founder with solid marketing skills will help you execute your target effectively and power growth. And if you are taking on an external co-founder, consider the legal and structural implications.

This means defining capital allocation, roles and responsibilities, and where the decisions are derived from. The definition of all of these aspects would forestall conflicts and misunderstandings later on and ensure smooth and supportive relationships.

External co-founders often come with their own networks of contacts and resources. This can lead to new strategic partners, or market opportunities that were previously inaccessible, significantly expanding the reach and potential of your business.

Investors

An entrepreneur needs investors who can provide capital, strategic advice, and important industry connections. The right investors put you on the fast track with your startup and open you up to a wide world of opportunities.

Investors provide the finances for you to scale up your business. It can be utilized for the development of products, expansion in markets, recruitment of talent, and other important business functions.

A dearth of funds is more often than not the difference between the failure and success of a startup. Wise investors bring a lot to the table through their network, experience, and understanding.

They can also provide valuable strategic advice on the operation of your business, market positioning, and growth strategies.

Their insight will guide you around the common pitfalls and enable you to make decisions that are wise and set you up for long-term success. Attracting investment from reputed investors can lend your startup respectability. It signals to the market, potential partners, and future investors that your business has been screened and is considered promising by professionals.

This way, it may become easier for you to get customers, talent, and additional investment. Investors are long-term stakeholders in your business success.

They provide ongoing support and resources to help your business grow and scale. This can be in the form of follow-on funding as well as advice for how to overcome current and upcoming challenges as your startup grows.

Service Providers

Beyond co-founders and investors, service providers are another critical component of the entrepreneurial network. These outsiders could offer specialized services that mean the most to running and scaling the startup.

Service providers will take care of the needs but often be too specialized or time-consuming; by outsourcing these tasks, your team can focus on strategic projects and critical business activities. Most of the time, service providers keep the newest tools and technologies that can be implemented to enhance your business operation.

For example, a tech provider may implement leading software solutions that better productivity and business workflows; an ad agency can use high-end analytics tools to optimize your marketing campaigns.

As your startup grows, it will have different needs for services. Service providers can offer scalable solutions that will grow with your business. Be it increased IT support as your user base grows or increasing marketing efforts for a big product launch, the service provider will be flexible to your changed requirements.

Sometimes, engagement with service providers is even cost-saving compared to building and retaining an in-house team for specific functions. This is extremely important for a start-up which desperately needs quality but cannot afford the costs associated with hiring personnel on a full-time basis for every function.

Now, service providers have a flexible pricing model that can fit different budget constraints. Many service providers specialize in regulatory compliance and risk management.

Legal and accounting service providers will ensure that your business complies with the laws and regulations, minimizing legal and financial risks. Delegating non-core functions to the service providers enables the companies to intensely concentrate on core activities like product development, customer acquisition, and strategic planning. This empowers your internal team to work with more efficiency and effectiveness towards business goals

Mentors and Advisors

Indeed, mentoring and advisory networks significantly contribute to making entrepreneurship successful because of the advice and support being given and the insights and vision into the business world passed down.

Entrepreneurs require as much wisdom and experience from their mentors and advisors for great success in the business world as the golden eagle does in its habitat.

These relationships are crucial to overcoming obstacles, capitalizing on opportunities, and driving growth. Finding and approaching potential mentors is a strategic approach, much like the golden eagle does in finding critical elements in its environment.

The very first thing an entrepreneur must do is to be very clear about identifying the needs to be in a position to know precisely what kind of guidance or knowledge could be most helpful.

To find potential business mentors who have found success in relevant fields, attend industry events, use online business networking sites such as LinkedIn, and search through professional associations.

Upon making the connection, entrepreneurs should state their intentions clearly, honor the mentor's time, and explain the desired expectations from the engagement, much like the golden eagle that engages its environment with precision and respect.

Building good relationships with mentors and advisors requires a lot of work and respect for one another, in the way that the golden eagle has consistent interactions and builds trust within their social structures.

Regular communication, with scheduled meetings, updates, or informal check-ins, maintains an engaged partner. Whether through thank-you notes, recognition in public, or finding ways to reciprocate when things get tough, find a way to show appreciation for them. Transparency about success and struggle builds trust and provides better odds of getting through. Acting on feedback will help show that their advice is valued and further reinforce the bond.

Industry Ecosystem

The industry ecosystem includes competitors, industry regulators, suppliers, and industry associations, among other stakeholders.

An industry ecosystem provides extensive benefits that cannot be done without the entrepreneur whose ambitions are to succeed. A presence in the industry ecosystem guarantees an entrepreneur current updates on trends, technology, and market dynamism.

By participating regularly in the conferences, trade shows, and seminars of the industry, we can keep an eye on the newly emerging trends and innovations. This will enable us to make strategic decisions from an informed position and take the leap against the competition.

Making alliances with another company in your business realm will be beneficial for both partners. The scope can range from joint ventures to research and development work. They can also access competitive intelligence that informs them of strategic planning by actively participating in the industry ecosystem.

Knowing your competitors' strengths and weaknesses and industry benchmarks can help identify their opportunities and threats.

This information is essential when positioning your start-up in the best way possible. It helps in building the reputation of your start-up.

This is going to add to your credibility and create many more business opportunities since you are going to be seen as one of the thought leaders or active contributors to the advancement within that industry. It also positions your startup as a significant player in the field.

Government and Public Sector

The business environment is also shaped by many levels of government—local, regional, and national—through their respective regulations, support programs, and policies. Government regulation sets the demand for a compliance process to be met while doing business by the law and ethical means.

These are tax laws, labor regulations, environmental standards, and requirements in respective industries. Indeed, governments often support start-ups with direct grants, loans, and subsidies—all useful to secure needed capital to initiate research and development, undertake expansion plans, and enable innovation.

Knowing and applying to them can significantly boost your financial resources without diluting equity. Several governments operate several programs for economic development and innovation.

Engaging with them unlocks networks, resources, and opportunities in line with the goals your startup has set forth. That said, another crucial role that the government plays is in protecting intellectual property through patents, trademarks, and copyrights.

Seizing IP rights is essential in protecting your innovation and gaining a competitive advantage. Engagement with relevant government agencies ensures that your IP is protected and enforced.

Challenges and Solutions in Networking

While building a strong network is vital for entrepreneurial success, it comes with its own set of difficulties and challenges. For most entrepreneurs, the first seems to be getting over their initial hesitance.

One must start by taking small steps to learn to approach networking with self-assurance; attending local events may help as it can prepare the ground in a more friendly environment.

However, preparation before an event by researching participants or developing talking points or questions might enable one to initiate conversations more easily. Furthermore, clear objectives of your goal help provide direction and reduce anxiety.

Time management is also a big struggle. Balancing time requires prioritization and a strategic plan about networking activities and other things. Focus on high-impact events most relevant to your goals and plan specific times to make your calendar consistent with networking activities.

Another way one can make networking more optimized in terms of time and less daunting in approach is by dovetailing it into other things already in place, like the time taken off at lunch breaks or business trips.

The bottom line for networking, though, is that one has to be authentic to truly connect and not just create more superficial relationships.

Authenticity attracts genuine connections, whereby others will appreciate you only if you are honest to yourself and share what you are passionate about and value deeply. Building authenticity includes paying attention to others, being sincerely interested in what they are talking about and their stories, and following up with people to develop a relationship past the first meeting.

Rejection is simply a part of the process; Rejection can be tough, but it's often a stepping stone to success. Handling it with grace and persistence shows resilience and determination.

Each rejection is an opportunity to learn and grow. Remember that rejection usually involves a mismatch of needs or timing and is seldom about personal failure.

Learning from rejection will hone your approach and render you more resilient. Stay positive and keep making new connections so that rejection does not derail your networking efforts but becomes a stepping stone for growth.

Building a Long-Term Network: The Golden Eagle Framework

Incorporating the golden eagle's adaptability and strategic vision, entrepreneurs can create a robust framework for building and maintaining a long-term network that fosters creativity and innovation.

Creativity often flourishes in diverse environments where different ideas and perspectives intersect. A network rich in varied expertise and experiences can inspire new ways of thinking and problem-solving.

Just as the golden eagle uses its environment to its advantage, entrepreneurs can tap into their networks to stimulate creativity, gaining insights and ideas that might not emerge in isolation. To create a framework that shows how creativity is correlated to the network, entrepreneurs can visualize their network as an ecosystem. In this ecosystem:

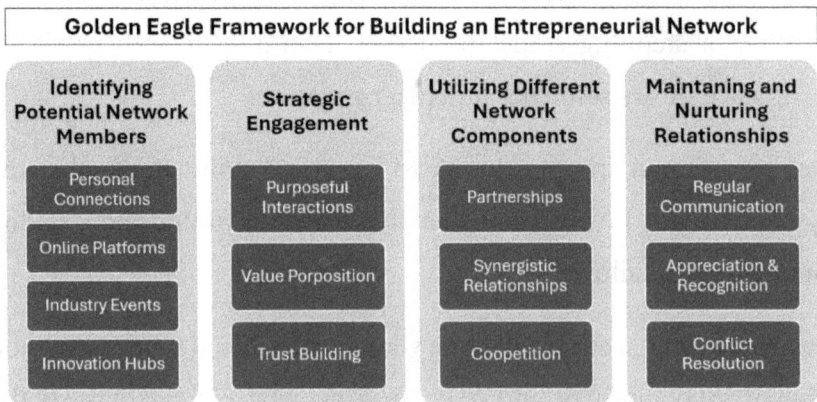

Golden Eagle Framework for Building an Entrepreneurial Network

Identifying Potential Network Members	Strategic Engagement	Utilizing Different Network Components	Maintaning and Nurturing Relationships
Personal Connections	Purposeful Interactions	Partnerships	Regular Communication
Online Platforms	Value Porposition	Synergistic Relationships	Appreciation & Recognition
Industry Events			
Innovation Hubs	Trust Building	Coopetition	Conflict Resolution

Identifying Potential Network Members

In the journey of building a successful entrepreneurial network, identifying and connecting with key members is paramount. This process can be effectively achieved through different ways:

Personal Connections

The role of personal networks in the world of entrepreneurship is very significant. Your current network of friends, family, and acquaintances could be a rich resource base for support, advice, and possibly even joint work. Quite often, the relationships built over the years carry a lot of hidden potential toward business.

A good and ideal way of benefiting from the power of people's networks is through informal gatherings. Hosting casual meet-ups, like a dinner party, a weekend barbecue, a coffee catch-up, can do that in a relaxed setting while dialogues of your entrepreneurial journey take place.

Here you are able to share the vision, ask for advice, and in a relaxed and friendly atmosphere examine possibilities for cooperation. For instance, over a casual dinner, you can pick up the fact that a friend has real insights about a market you are trying to get into or knows someone who can be a key partner or investor.

Online Platforms

In this digital age, online platforms have changed the way entrepreneurs network. Nowadays, professional networking websites, industry-related forums, and social media networking have become vital means of building and maintaining contacts.

The largest professional networking site, LinkedIn allows entrepreneurs to join groups of industries, participate in discussions, and follow companies and influencers.

Within LinkedIn groups related to particular industries, such as the renewable energy sector, entrepreneurs can connect with industry experts, hold meaningful discussions, or reach out to partners and investors.

Industry-specific forums and communities are all over the web, from places like Reddit and Quora to specialized industry communities, where one can both ask questions and demonstrate knowledge of a subject.

Being an AI startup founder and involved in the discussion on AI-related subreddits or on Stack Overflow may help him connect with other passionate people over artificial intelligence and thus open avenues for potential collaboration in the future.

Some of these networking opportunities are through social media platforms, such as X, Instagram, and Facebook. A strong current through industry news, trends, and conversations, using relevant hashtags and following key figures, helps entrepreneurs keep updated and network with other like-minded individuals. For instance, the hashtag #FinTech on X allows following the discussion around new developments that one can interact with key players in the financial technology space.

Industry Events

Industry events are the prime events for networking, where professional experts, innovators, and leaders in the said field congregate to share their insights, showcase advancements, and, more importantly, explore opportunities. These range from extensive conferences and trade shows to intimate workshops and seminars.

Attend events such as industry-specific conferences and trade shows, where one gets a chance to meet the major players who are shaping the industry.

For example, it is a great opportunity for tech entrepreneurs to witness such major events as CES (Consumer Electronics Show), in which promising start-ups and investors looking for new disruptive ideas take part, besides major technology companies. Such events turn out to be a market of living ideas, through informal conversation which can lead to a strategic partnership or investment opportunities.

Workshops and seminars are a potential platform for learning and interacting in a specialized setting. Entrepreneurs can then take the smaller settings to have deep discussions on such topics that help them in skill development and networking. For example, joining a marketing workshop can link you up with various marketing experts and potential clients who may be interested in innovative strategies and tools.

Networking mixers and social events organized alongside these conferences and workshops make for a relaxed environment, where participation becomes much higher. For example, an evening mixer at a biotech conference can lead to casual yet fruitful conversations with potential collaborators or investors, fostering relationships that go way beyond the event itself.

Innovation Hub

Innovation hubs in the form of incubators, accelerators, and co-working spaces are breeding grounds for creativity, collaboration, and growth. They provide resources, mentorship, and networking required for entrepreneurial success.

Incubators and accelerators are structured programs that support startups in funding, mentorship, and the broader network of successful entrepreneurs and investors. For example, Y Combinator does not only provide financial help but also offers advice and connections invaluable for technology startups on their way to new heights.

Spaces such as co-working spaces, for example, WeWork and Impact Hub, are poised to bring together entrepreneurs, freelancers, and small businesses. Spaces usually hold networking events, workshops, and collaborative projects where interaction and strategic partnerships among like-minded people can take place.

Innovation and research centers, like the MIT Media Lab, facilitate cooperation between startups and established companies on state-of-the-art research projects. Engaging with such centers gives entrepreneurs access to advanced technology and the expertise needed to stay ahead in their respective fields.

Strategic Engagement

Strategic engagement characterizes the kind of deliberate networking that works well in entrepreneurship. It is all about relating to others in a way that will have an explicit purpose and, at the same time, ensuring that through each interaction, the value gets added, which concretely builds the network.

Purposeful Interactions

Purposeful interactions are key to strategic engagement. Enter every conversation and meeting with something that you want to achieve. Be it exploring a potential partnership, seeking advice on how best to secure an investment, or having a given objective, clearly defining your purpose keeps the interaction meaningful and of mutual benefit.

Think of being at a conference that gives you the chance to get personal one-on-one time with industry leaders and possible collaborators. They're not looking for the exchange of small talk, though, but of interaction with some purpose at its core. You set it up by learning about the people you want to meet and how they help to realize your objectives. When you clearly articulate your goals and listen for theirs, a dialogue results that is meaningful and impactful: purposeful interactions. It isn't about taking; it is about creating value for both parties.

Value Proposition

Central to strategic engagement is the ability to clearly communicate your value proposition. In every interaction, it is crucial to articulate what you bring to the table and how it benefits the other party.

Whether you are proposing a partnership, seeking investment, or collaborating on a project, a compelling value proposition can set you apart and capture interest. It should always be framed as a win-win situation, showing how both parties will benefit from the collaboration.

Consider a startup founder pitching to potential investors. Beyond showcasing the innovative aspects of the product, the founder must clearly articulate how the investment will yield substantial returns.

This involves presenting the financial projections and highlighting the unique strengths of the team, the market opportunity, and the strategic advantages of the business model. By effectively communicating how the partnership will be mutually beneficial, you make it easier for others to see the value in engaging with you.

Trust Building

At the core of every good network is trust. Building and maintaining trust takes reliability, consistency, and transparent communication. You need to earn trust by acts showing your integrity and commitment over time.

Reliability means that you follow through in what you have agreed to do. If you committed to a date for delivering a report or to a meeting, meet those commitments.

Consistent actions support your reliability: you demonstrate yourself to be dependable and trustworthy. Transparent communication is also important. Openness about intentions, challenges, and progress helps create an atmosphere of honesty and mutual respect.

Utilizing Different Network Components

A strong entrepreneurial network is built of not just the right people to get involved with, but also using the relationships between the people to achieve maximum potential in business. Here are three critical ways to utilize different network components:

Partnerships

Strategic partnerships can be one of the most powerful channels to market, where the opportunity to co-develop products is playing to each other's strengths. Just align with companies that complement yours and vice versa in order to create synergy, which results in mutual growth and success.

For example, a tech startup may collaborate with a large software company. The startup will be able to enter new markets through the distribution channels that have long been developed by the software company, while the latter gains access to innovative technology developed by the startup.

Synergistic Relationships

Synergistic Relationships are a kind of relationship between all parties, who make the effort to provide a result. In such relationships, the strengths and capabilities of each party will lead to improve the outcome and achieve the goals. Mutual benefit, high effectivity and collaborative innovation are considered as the key characteristics of synergistic relationships.

An example could be that a university's engineering department and a manufacturing company form a synergistic relationship to advance research in sustainable manufacturing practices.

Coopetition

Your competitors or partial competitors are also among those who can make long term relationships with you. There are many alliances in the business world that happen between competitors and so many joint projects have been made through the blend of competition and cooperation.

This concept for entrepreneurs means that as an entrepreneur, you need to build some sort of relationships with your competitors that could benefit both parties. This is a ticky action though and should be done with careful consideration and strategic approach.

For example, Samsung and Apple compete while Samsung supplies Apple with OLED screens and some high quality parts. This partnership is a win-win that makes a revenue stream for Samsung while Apple is satisfied with the high quality parts that are supplied by Samsung.

Netflix and Amazon also are two giant competitors in the streaming world, while Netflix relies on AWS to host their contents; Amazon makes money from the competitor and Netflix has a reliable cloud solution. Netflix demands also help AWS to innovate and improve the cloud infrastructure.

Maintaining and Nurturing Relationships

In the world of entrepreneurship, making connections is just the start. The true strength of a network comes from how well these relationships are maintained and nurtured over time.

Regular Communication

Regular communication is the cornerstone of a thriving network. Keeping in touch with your contacts through updates, meetings, and check-ins ensures that your relationships remain active and engaged.

This consistent interaction helps keep everyone informed about your progress, upcoming projects, and any challenges you might be facing. This could mean something as simple as sending out a monthly newsletter to your network to update them on your company's developments, achievements, and future plans.

Regular meetings in person or virtually allow for an opportunity to discuss ongoing projects, to share ideas, and to reinforce connections.

Appreciation & Recognition

Showing appreciation and recognition for contributions is vital for nurturing strong and positive relationships. Value your network members by developing a program to appreciate and recognize them: thank-you notes, public acknowledgments, and recognition programs all mean a lot.

For example, after a successful project collaboration, send out personalized thank-you notes to everyone that took part; it shows appreciation and a reminder that their work really matters.

Publicly acknowledge the contributions of your partners or team members in meetings, newsletters, or on social media to highlight the achievements and foster pride and loyalty. Formal recognition programs, such as "Partner of the Month" awards, should foster continued collaboration by making such collaborations a culture of appreciation.

Conflict Resolution

Conflict is very natural to come by in every relationship, but it is how you react to it that will determine if you can maintain trust and harmony within your network.

Conflict resolution that is timely and empathic will lead to issue resolution before things get out of hand and will not make any party feel less heard or respected.

When conflicts arise, approach them with an open mind and a willingness to understand the perspectives of all involved. Transparent and honest communication can help to clear any misunderstandings between both parties and find common ground.

For example, where there is a dispute with a business partner regarding the timelines of a project, fronting the issue in an open and collaborative manner can help to arrive at a mutually acceptable solution, further building the partnership.

In other words, empathy is critical to conflict resolution because it shows the other person that you really want to have a relationship in which you two treat each other in a respectful, supportive, and healthy way.

The Golden Eagle Entrepreneur and Networking

Networking is a fundamental aspect of entrepreneurial success, akin to the golden eagle's survival in the wild. The golden eagle excels by understanding its environment, making strategic connections, and adapting to changing conditions. Entrepreneurs can adopt a similar approach to build and leverage their networks effectively.

To navigate the business landscape, entrepreneurs should identify key individuals and groups within their industry and community, just as the golden eagle recognizes crucial elements in its surroundings.

This involves pinpointing potential partners, competitors, mentors, and influencers. Understanding the dynamics of these relationships is essential; observing how different members interact, understanding their needs and goals, and determining how to add value to these connections are vital steps.

CHAPTER TEN

CONTINGENCY, ADAPTABILITY, AND RESILIENCE

"The moment you accept what troubles you've been given, the door will open." - **Rumi**

Much like the golden eagle that navigates varying terrains and unpredictable weather, successful entrepreneurs must master the art of contingency planning, adaptability, and resilience.

In this chapter, we'll discuss the core attributes that make the golden eagle entrepreneur thrive in the face of uncertainty.

We will explore how to develop strategies that ensure flexibility, build resilience to overcome setbacks, and harness the power of adaptability to maximize opportunities.

By understanding these principles, entrepreneurs can fortify their startups against the challenges of the business world.

T he path to success in entrepreneurship is rarely a straight line. Navigating this journey requires more than just an innovative idea and a visionary mindset; it demands a high tolerance for ambiguity. Tolerance for ambiguity allows entrepreneurs to stay composed and make decisions despite uncertain and unpredictable circumstances. This, coupled with contingency planning, prepares them for unforeseen events, ensuring they have a roadmap when things don't go as planned. Adaptability is the skill to pivot and adjust strategies in response to changing circumstances, while resilience is the inner strength to bounce back from setbacks and continue moving forward. Together, these qualities form the backbone of a successful entrepreneurial endeavor, enabling leaders to thrive in an ever-evolving landscape.

Contingency in Entrepreneurship

Contingency refers to the preparation for unforeseen events or emergencies that might happen in the future and we don't know what they are.

The root cause of the need for a contingency plan in startups and even larger corporations is the unknown future and high degree of ambiguity and uncertainty.

Uncertainty and Ambiguity

As mentioned, uncertainty and ambiguity can either paralyze or propel an aspiring entrepreneur. So, in the face of these challenges, we turn our focus to the next crucial concepts:

Building the Unknown Future

Building the future is an endeavor that involves stepping into the realm of the unknown, daring to imagine possibilities that have yet to be realized.

This journey requires vision, creativity, and a willingness to embrace uncertainty. At the heart of this process lies the matrix of the known and unknown, a conceptual framework that helps innovators navigate the complexities of envisioning and creating the future.

Donald Rumsfeld Matrix

The matrix of knowns and unknowns, often referred to as the "known unknowns" matrix, was designed and used by Donald Rumsfeld, the former U.S. Secretary of Defense.

He popularized and articulated the concept in a 2002 press briefing, although the idea has been used in risk management and strategic planning in various forms prior to that.

A matrix of knowns-unknowns helps in categorizing different types of uncertainties and can guide entrepreneurs in navigating complex and unpredictable environments.

Known-Knowns	Known-Unknowns
Things we are aware of and understand. Examples: current market conditions, existing competition.	Things we are aware of but don't understand fully. Examples: potential market trends, upcoming regulatory changes.
Unknown-Knowns	Unknown-Unknowns
Things we are not aware of that we know. Examples: untapped potential within the company, latent skills or resources.	Things we are not aware of and don't understand. Examples: future disruptive technologies, unforeseen global events.

Known-Knowns: These are factors or information that an entrepreneur is fully aware of and understands. For example, they might know the current market size, their competitors, and their business capabilities.

Known-Unknowns: These are factors or information that an entrepreneur is aware of but does not fully understand. For example, they might know that a new regulation is coming but are uncertain about its impact on their business.

Unknown-Knowns: These are factors or information that an entrepreneur possesses but is not aware of. For example, they might have a team member with a skill set that can be pivotal for a new project but have not recognized or utilized it yet.

Unknown-Unknowns: These are factors or information that an entrepreneur is neither aware of nor understands. This quadrant represents the greatest uncertainty and risk, such as future disruptive technologies or unexpected global events.

The Golden Eagle Entrepreneur and Rumsfeld Matrix

The golden eagle entrepreneur is distinguished by their exceptional ability to navigate and capitalize on opportunities across all four quadrants of the matrix. This unique prowess stems from a combination of visionary thinking, resilience, adaptability, strategic foresight, and an innovative mindset.

The golden eagle entrepreneur possesses an extraordinary capacity for visionary thinking. This ability allows them to foresee and shape future trends, particularly in the realm of Known Unknowns. They can identify emerging market demands and technological advancements before they fully materialize, positioning their ventures to exploit these opportunities effectively. Moreover, their visionary approach enables them to pioneer groundbreaking innovations within the Unknown Unknowns quadrant. By embracing the ambiguity of the future, they are often at the forefront of disruptive innovations that redefine entire industries.

This matrix can be the base for the contingency planning for founders. To build the unknown future, innovators must actively engage with each quadrant of the matrix, especially the unknown-unknowns. This requires a mindset that is open to exploration, experimentation, and the willingness to accept and learn from failure.

Adaptability and Entrepreneurship

Adaptability in entrepreneurship involves flexibility in thinking, continuous learning, and responsive decision-making.

The golden eagle entrepreneurs are active as they take action to build the future, but as they grow, they should compromise their activeness and add some degree of reactiveness to their leadership style in order to be adaptable.

There might be a need to pivot in their business models, explore new markets, and adopt innovative technologies despite unclear circumstances.

Adaptability and entrepreneurship are closely intertwined concepts that significantly contribute to the sustainability of businesses in a rapidly changing world. Adaptability refers to the ability to adjust to new conditions, respond to changes, and modify strategies in response to evolving circumstances.

Adaptability, VUCA, and OODA Loop

The business environment is in some cases considered with characteristics of Volatility, Uncertainty, Complexity, and Ambiguity (VUCA). As mentioned earlier, uncertainty and ambiguity are tied with entrepreneurship, but not every startup in the beginning faces complexity and volatility.

As the startup grows, they may face volatile conditions arising from market changes, regulation changes, or rapid technology changes and complexity will show sooner or later. In such conditions, being aware and prepared helps the founders manage VUCA conditions better.

The OODA Loop is a decision-making framework developed by military strategist Colonel John Boyd. It stands for *Observe, Orient,*

Decide, and Act and is an iterative process. This loop is designed to help individuals and organizations make quick and effective decisions in complex and rapidly changing environments. Here are the iterative steps of OODA Loop:

Observe: Collect current information from as many sources as possible.

Orient: Analyze this information to form an understanding.

Decide: Determine the best course of action based on the analysis.

Act: Implement the chosen decision and observe the results.

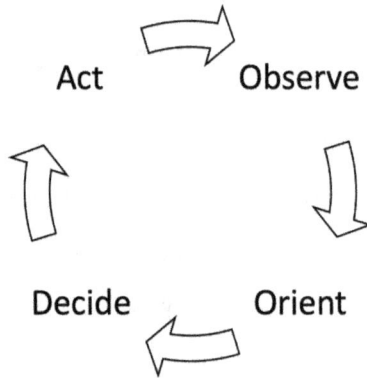

Act   Observe

Decide   Orient

As mentioned, the golden eagle hunting strategies involve great adaptability. When an attempt to capture the prey is done, there is a high degree of uncertainty and ambiguity that the eagle involves as it does not know how the next maneuver of the prey to escape is.

The golden eagle acts, observes, and when a maneuver from prey happens orients the course and makes the next decision along the way.

The adaptability of the golden eagle entrepreneur is also the same; they will start and along the way keep an eye on the internal and external factors and orient. The Lean startup concept of Eric Ries and Customer Development from Steve Bank also suggest the same. The agility of startups requires adaptability of a golden eagle entrepreneur.

"But man is not made for defeat, man can be destroyed but not be defeated." - Ernest Hemingway, "The old man and the sea".

Resilience and Entrepreneurship

Nowadays, most people talk about resilience when it comes to entrepreneurial challenges. Founders face a variety of problems, obstacles, and challenges in their entrepreneurial journey and should try to stay focused in spite of all adversaries and problems.

Resilience is not just the ability to tolerate setbacks and challenges; but it is also the capacity to recover quickly from difficulties and keep focus. For founders, resilience is an essential trait that fuels perseverance, creativity, and sustainability as an entrepreneur.

Founders will surely face times that they get disappointed and setbacks and failures have a tremendous mental impact on them. It is very hard to make decisions and sometimes the stress and mental pressure paralyzes the entrepreneur. Here is the author's experience:

"I was too bold at the time of Lar Fishing in Sweden and once I made a big purchase of inventory with all I had in my pocket expecting to have some sales that didn't happen. When you make an expectation and

forecast and things don't go as expected, you will get the most devastating surprise as a founder.

The money in my bank was enough for one grocery shopping and bills to pay were due in three days. I panicked and could not do anything. The stress paralyzed me and depression was on the way.

Suddenly, I got up, I filled my backpack with fishing flies and other products that I imported directly myself and knew the price was good and went to the local fishing store who was sort of my competitor. I just went to the store owner and asked them to buy, telling them that I imported them and I can give them a good deal.

Surprisingly, I came back with an empty backpack. That was the time I learned about 'coopetition' (cooperation with your competitor) and how it can work but the lesson for me still works. In order to manage resilience and overcome failure and setbacks, you should do something and act.

Staying passive and being paralyzed is the source of 95% of the stress and mental problems, but as an entrepreneur you should do something and as long as you do something, there will be less mental problems."

Resilience involves a positive mindset that views failures and setbacks as learning opportunities rather than obstacles and defeat. Entrepreneurs who possess resilience understand that failure is an inevitable part of the journey.

They embrace these challenges, extracting valuable lessons that inform future strategies and decisions. This ability to learn and grow from adversity is crucial in an environment where uncertainty and risk are your constant companions.

A resilient entrepreneur maintains a clear vision and unwavering determination, even when confronted with significant challenges.

This vision acts as a guiding star, helping them navigate through turbulent times with focus and clarity. It is this steadfast commitment to their goals that enables them to stay the course and persist when others might give up. The ability to maintain such focus and determination is a defining characteristic of resilient leaders.

Adaptability also helps entrepreneurs when it comes to resilience—the ability to pivot and adjust strategies in response to changing circumstances can help them make better moves in tough conditions.

In a volatile market, entrepreneurs must be flexible and open to change, continuously refining their approach to meet new demands.

This adaptability is powered by resilience, allowing entrepreneurs to view change, not as a threat but as an opportunity for innovation and growth. By staying agile, they can respond to market shifts swiftly and effectively, ensuring their business remains relevant and competitive.

As an example, with the rise of generative AI, some entrepreneurs might see this threat as A new AI product is taking their market share, and in such tough conditions some might get demotivated and get ready to give up.

A golden eagle entrepreneur, on the other hand, is flexible, agile, and resilient enough to add generative AI to their product.

Another critical aspect of resilience is emotional intelligence. Entrepreneurs who are emotionally intelligent can manage their emotions, maintain perspective, and stay composed under pressure.

This emotional stability helps them make better decisions, build strong relationships, and lead their teams effectively through challenging times. It also fosters a positive work environment, encouraging collaboration and collective problem-solving.

This is where most nascent entrepreneurs make their biggest mistakes, when they make decisions based on impulsive reactions and without patience.

Resilient entrepreneurs also build strong support networks. They understand the value of surrounding themselves with mentors, advisors, and peers who can offer guidance, support, and encouragement.

These networks give them the ability to bounce back effectively and stay focused. Collaborative relationships and a sense of community can significantly enhance an entrepreneur's resilience, providing the resources and moral support needed to persevere.

The Golden Eagle Entrepreneur

The golden eagle entrepreneur is defined by their ability to navigate uncertainty with grace and determination. Embracing contingency, they prepare for unexpected challenges by developing comprehensive backup plans and maintaining a flexible approach. This readiness allows them to pivot quickly and effectively when faced with unforeseen circumstances, turning potential setbacks into opportunities for growth.

Their ability to anticipate and plan for multiple scenarios ensures that they are never caught off guard, embodying the foresight and strategic planning that are hallmarks of successful entrepreneurship.

Adaptability is another cornerstone of the Golden Eagle Entrepreneur's success. Like the majestic bird that adapts to varying environments and hunting conditions, these entrepreneurs remain agile and responsive to changing market dynamics.

The OODA Loop (Observe, Orient, Decide, Act) framework is integral to their adaptability, helping them make swift and informed decisions in a VUCA (Volatile, Uncertain, Complex, Ambiguous) environment. They continuously scan the business landscape, absorb new information, and adjust their strategies accordingly.

This dynamic approach enables them to stay ahead of the curve, innovate consistently, and meet the evolving needs of their customers. Their willingness to embrace change and learn from it allows them to thrive in an ever-shifting business environment.

Resilience ties together the traits of contingency planning and adaptability, forming the backbone of the Golden Eagle Entrepreneur's journey.

Resilience is about more than just bouncing back from failure; it is about learning from experiences, continuously improving, and maintaining an unwavering belief in one's vision. This belief is not just a passive hope but an active, driving force that propels them forward, even when the skies are stormy.

Therefore, the golden eagle entrepreneur faces challenges head-on, leveraging their network, focus, and bold initiatives to overcome obstacles and achieve their goals. Their resilience is reflected in their ability to maintain a positive outlook, persist through adversity, and emerge stronger and wiser from each challenge. This tenacity and grit are what ultimately propel them to soar to new heights of entrepreneurial success.

CHAPTER ELEVEN

FAILURE & THE FEEDBACK LOOP

"The wound is the place where the light enters you." **— Rumi**

Every hunt is a unique adventure, and let's face it, not every attempt is going to be a success. But that's okay! The golden eagle doesn't get discouraged by failure but learns from it! With the help of a feedback loop, it tries to do better next time.

The golden eagle entrepreneurs leverage feedback loops to refine their strategies, products, and operations. This chapter explores the critical role of failure and the essence of feedback loops in the entrepreneurial journey, providing insights into how to achieve sustainable growth by embracing failure and using a feedback loop.

Failure in Entrepreneurship

As discussed, entrepreneurship involves inevitable risks and risks bring possible setbacks and failures as the result of a bold and innovative action during an entrepreneur's journey.

We also talked about the resilience of entrepreneurs and how to deal with tough conditions and failures they might face along the way. But entrepreneurial failure is inevitable, as the practice of entrepreneurship involves risk and innovation. In the previous chapter, we mentioned how to manage resilience caused by failure and setbacks.

The golden eagle entrepreneurs, however, are not afraid of failure and build a bridge of success based on the failure.

Understanding Feedback Loop

Feedback loop is the process in which the output of a system is fed back into the input of that system again, so that it can provide better output results by refining the process and adapting to new changes that are required. This is the base of a continuous improvement that could lead to better decisions, growth, sustainability, and innovations.

There are two types of feedback that startup founders have to take into consideration: external feedback and internal feedback. The internal feedbacks are inside the team and the organization, and the external feedbacks are the ones that are coming from outside of your organization.

Internal Feedbacks

Team Feedback

The first feedback you need to get is the team you are working with. A founder should implement plenty of different mechanisms and tools to ensure continuous feedback is taken from the team. Here are the major ways of getting feedback from your team:

- Regular all-hands meetings that could be bi-weekly or monthly to listen to everyone and discuss challenges or celebrate success.
- One-on-one meeting among team members in particular when a personalized feedback is needed
- A comprehensive, well organized 360 degree feedback that everyone receives from peers, subordinates, and managers could help everyone better understand their performance and the area of their improvements
- Regular retrospectives after a project or finishing a sprint, which is one of the parts of the agile methodology, allows the team to learn and share about what went well and what didn't

When we want to build a feedback culture in our organization, we have to start inside first with our team. Leaders who are not able to listen to their employees and manage to inject the feedback culture inside, will never listen to those more important external feedbacks that comes from *customers, market,* and *key partners.*

Operational Feedback

Operational feedback is a continuous process where Key Performance Indicators (KPIs) are measured and based on the results the execution of tasks is analyzed.

This type of feedback focuses on the day-to-day operations and processes that are essential for achieving the organization's goals and objectives that are set before. By implementing a robust operational feedback system, startups and scaleups can ensure that their operations are running smoothly, identify areas for improvement, and make necessary adjustments in real-time.

There are different mechanisms for operational feedback that are tied with other types of feedback such as customer feedback and team feedback. Each startup should come up with their own KPI and their own performance management system and make sure to measure and review their KPIs on a regular basis.

External Feedbacks

Customer Feedback

Among external feedback, *customer feedback* matters the most, as they are the ones who will use and pay for your product or services. Customer feedback is also crucial for you to find out your product-market fit and grow. There are different ways to get feedback from customers.

The following table shows the most effective customer feedback methods.

Feedback Method	Type	Advantages
Surveys	Quantitative	Scalable, easy to distribute
Interviews	Qualitative	In-depth insights, personal interaction
Focus Groups	Qualitative	Detailed discussions, group dynamics
Feedback Forms	Qualitative	Convenient, continuous input
Social Media Monitoring	Mixed	Real-time tracking, broad reach
Review Websites	Qualitative	Public opinions, credibility building
Ratings and Reviews	Quantitative	Product-specific feedback, trust factor
Net Promoter Score (NPS)	Quantitative	Measures customer loyalty, simple metric
Support Tickets and Live Chat	Qualitative	Issue-specific insights, direct communication
User Testing	Qualitative	Usability insights, user behavior
Advisory Boards and Workshops	Qualitative	Strategic insights, collaborative feedback
Post-Purchase Feedback	Quantitative	Immediate experience feedback
Behavioral Analytics	Quantitative	Data-driven user behavior insights

Market Feedback

Market feedback is the process of gathering and analyzing information from the market to understand customer preferences, market trends, and competitive dynamics.

This feedback is used for making informed business decisions, improving products or services, and identifying new opportunities. Market feedback methods include all customer feedback plus conducting market and industry analysis, market trend analysis, competitor analysis, etc.

Key Partners Feedback

Feedback from key partners and stakeholders, such as mentors, suppliers, and investors, is crucial for startups, offering strategic guidance, operational insights, and opportunities for innovation.

Regularly engaging with these partners through meetings, surveys, and focus groups can provide valuable perspectives on key factors such as market trends, supply chain efficiency, and financial performance.

Utilizing this feedback helps startups refine their strategies, improve operations, and foster innovation, ultimately leading to more robust and mutually beneficial partnerships. This collaborative approach ensures that startups remain agile, responsive, and well-positioned to navigate challenges and seize new opportunities.

Openness To Feedback

In her seminal book "Mindset," psychologist Carol Dweck introduces the concepts of fixed and growth mindsets. A fixed mindset is the belief that abilities and intelligence are static and unchangeable. Individuals with this mindset often think their talents and skills are innate gifts that cannot be significantly developed over time. This perspective contrasts sharply with a growth mindset, which embraces learning, development, and the idea that intelligence and abilities can be cultivated through effort and persistence.

People with a fixed mindset tend to exhibit high resistance to change, which can lead to stagnation in their business ventures.

They may be more inclined to stick to familiar methods and strategies, avoiding new challenges and opportunities that require them to step out of their comfort zones. This resistance to change can be particularly detrimental in the fast-paced and dynamic world of entrepreneurship, where adaptability and innovation are key to staying competitive and relevant.

Moreover, people with a fixed mindset are not open to feedback. Entrepreneurs with this mindset might view criticism as a personal attack rather than an opportunity for growth.

As a result, they may isolate themselves from valuable insights and perspectives that could help improve their business. This isolation further limits their potential for innovation and growth, as they are less likely to iterate and refine their ideas based on external input. As the author explains:

"As a mentor, I always asked founders to hear what I suggest, but do whatever their guts tell them and follow their intuition. Some entrepreneurs are obsessed with their idea, which is not necessarily their startup idea and they never even hear you. They have a closed mind and only believe in what they want to do. These founders always learn things the hard way; by failing and making mistakes."

Ego also plays an important role when it comes to being open to feedback. Founders with too much ego are the ones who destroy the team and block the performance improvement and growth of their business. There is however, a positive side of ego that is giving the founders confidence and drive.

To overcome the limitations of a fixed mindset, founders should work on developing a growth mindset. This involves embracing challenges, viewing failures as learning opportunities, and understanding that abilities can be developed through effort and perseverance. By fostering a growth mindset, entrepreneurs can become more open to experimentation, more resilient in the face of setbacks, and more willing to adapt to changing circumstances. As the Persian poet, Rumi says:

"Yesterday I was clever, so I wanted to change the world. Today I am wise, so I am changing myself." – Rumi

Balancing the Feedback Loop

A feedback loop is beyond learning from failure and mistakes; it fosters the growth environment in your startup, but as the author describes in one of his experiences, there should be a balanced approach:

"Legal issues, especially the ones that are taking you to the court for the first time as a founder can destroy the founder. For me it was tough, when I faced the legal case at Lar Fishing and in Sweden and I had not seen a court or a lawsuit in my life before that.

When I met my business coach Göran Nilsson I heard something that I will never forget: 'In such cases everyone is going to tell you that you should learn from your experiences. That's true, but DO NOT LEARN TOO MUCH!'."

There should be balance when you are going to listen to your team, your mentor, your customers, or learning from failure. Failure should be a learning incident and you need to have a growth mindset and be open to feedback, but if you listen to every feedback, new problems will arise. There is only one formula for this balance between being focused and determined and open to feedback.

The entrepreneur should listen to their guts and intuition more than anything when making decisions after being open to hear feedback and being aware of what's going on.

The golden eagle entrepreneur has different intelligence, mindsets and capabilities, and the biggest innovations in the world came from those innovations and entrepreneurial practices that were started and led mostly by the intuition and brain of the golden eagle entrepreneurs.

The golden eagle entrepreneur is flexible enough to change the course, is open to hear the feedback and uses the feedback loop in a constructive way, but he also has some degree of ego and decisiveness that gives him confidence to initiate and take risks. The golden eagle entrepreneur sees risks, accepts risks, is not afraid of failure, and does not learn too much.

CHAPTER TWELVE

THE GOLDEN EAGLE INTELLIGENCE

"Wisdom and intelligence are the wings that will carry you far and high, to lands unseen and dreams undreamed"- **Ferdowsi**

"The Golden Eagle Intelligence" is all about the decision making of the golden eagle entrepreneur. The golden eagle intelligence reveals what's inside the golden eagle entrepreneurs' brain and how they think and decide.

In the previous chapters, we discovered exploring opportunities, Future foresight, innovation, risk appetite, focus, bold initiations, network, contingency, adaptability, resilience, failure, and feedback loop.

In this chapter, we'll connect the dots and see how these elements are orchestrated, making the golden eagle entrepreneur soar higher than the others.

Intelligence

We run into the term *intelligence* in our life frequently. IQ or intelligence quotient is a test to find how intelligent a person is, EQ or Emotional Intelligence represents how emotional intelligent the person is; people in business hear about business intelligence tools, nowadays everyone around us is talking about Artificial Intelligence (AI) and the world politicians main decisions are based on the intelligence they get from their intelligence services.

Peter Earnest, a former CIA officer and his co-author Maryann Karinch in their book *Business Confidential* state that intelligence is the product of *collection* and *analysis* of information, which in the spy community is collected in a way that we do consider unethical, illegal, and improper in the entrepreneurs and business community, but we are not restricted to learn from the applicable and ethical parts of their work in the business world.

Earnest regards exclusivity as the core value of intelligence: *"if the collected information is analyzed, but is also available via purchase or other ways, it is not intelligence"*.

Entrepreneurial Intelligence

Entrepreneurial Intelligence (EI) is the result of the information processing framework we discussed in chapter three. EI is gained by individuals in their life's journey and there are many factors that involve entrepreneurial intelligence.

Collection Information

The collection of data, information, and knowledge is the base for reaching entrepreneurial intelligence. Every entrepreneur needs to collect information and data through different sources and acquire knowledge and experience through some other sources that act as the raw material for analysis. Before getting into the collection phase, you have to define your questions.

In the book "Elon Musk", Ashley Vance refers to this as one of the behaviors of Elon Musk. *"Musk believes that one of the hardest things is to find out what questions to ask and once you know what question to ask, the answer is relatively easy."* Knowing what questions to ask and what you want gives purpose to the collection phase.

Once you know the questions, you have the next milestone, which is answering that through collection, analysis, and conclusion. There are a couple of sources for collecting data, information, or knowledge.

When dealing with data and information there are a couple of collection sources. *Humans are* primarily the main source of your information you need. Interview is the most popular method for collecting information.

Interview is usually a formal session with an expert in the field in order to get a better understanding of the opportunity or related parts of it. In some cases these field experts have deep experiences and they may also possess wisdom.

Even when we are talking to people of wisdom, you should not expect to get intelligence or wisdom from them. This might seem irrational to you, but remember that you are going to be the entrepreneur who does the analysis and anything you get from a wise or intelligent person is not more than a knowledge to you.

It is your own personal background, your knowledge, your domain expertise, and your brain that makes a new intelligence out of all input and eventually makes the decision.

When collecting information directly from people, remember that this is not required to be formal and structured.

Also we also mentioned in chapter nine, sharing and exchange of information is an integral part of acquiring new information as it is a two-way street.

Deborah Perry Piscione, in her book "Secrets of Silicon Valley " maintains that one of the main reasons that Silicon Valley is much more successful than the rest of the world's startup ecosystems is *the sharing culture*.

The interesting part is that she came from Washington DC, and from a political background and made a comparison regarding information sharing.

Unlike Washington DC, in which everyone wants to keep their information as confidential as possible, in the Valley everyone trusts others and shares information, sometimes even with competitors. This is the real difference between Silicon Valley and other parts of the world.

Some information sources are available for all people and you can collect your information through open sources. These collection sources might be free or you may have to pay for them sometimes.

News media, libraries, journals, trade shows, published statistics, market insights, reports and attending conferences and trade shows.

Open sources are not purposeful information collection like the ones we are used to seeing in market research. Reading books, and developing your own knowledge is the most important part of acquiring open sources of information and knowledge.

Analysis and Compression

Analysis of what you have collected is an important process, in which the human factor is absolutely dominant. No matter what sort of data, information, or know-how you possess and collect, the final analysis is completely the core of the human brain that leads to the act of entrepreneurship.

You may use the support of computer and artificial intelligence, which is inevitable, but the final decision of an entrepreneur is made inside the entrepreneur's brain.

The outburst of the latest investments on Artificial Intelligence (AI), machine learning, text mining, robots and accurate sensors, and other high-tech tools might make most people think that information processing and making the decision can be done by the machine.

The human brain was also questioned by such entrepreneurs as Elon Musk in the past, but even Elon Musk who is the icon of future technology development in the world realized his mistake.

In the past, Musk was insisting that the human brain is incapable of dealing with some tasks, and in particular he was saying that self-driving cars save lives, just because they make the decisions and analysis instead of the human brain. He also made too much automation, but after missing some deadlines in the production of Tesla he tweeted:

"Yes, excessive automation at Tesla was a mistake. To be precise, my mistake. Humans are underrated."

As mentioned earlier, there are two types of information processing: *algorithmic* and *heuristic*.

Algorithmic analysis involves dealing with patterns and is done by formulating available information. In this method, the human brain looks for frameworks, models, patterns, or systematic events, and formulates a pattern from them.

Some patterns can be recognized easily, while some other patterns are more complicated and need more analysis. A drop in sales is obvious from numbers if you have your monthly revenue in front of you, but reaching a new business model for your startup in the Fintech industry or investing in a startup requires much more time and analysis.

Heuristic analysis is based on trial and error, your prior knowledge and some of your previous experiences of anything. Heuristic is usually about your own biases and is a subjective analysis.

Some people are against biases when it comes to making decisions, which might be true in other sectors than entrepreneurship.

Entrepreneurial intelligence is also about the entrepreneur's biases and their subjective decision making. The heuristic analysis is connected to the algorithmic and while there is also a connection between the two types, the entrepreneur's mind makes a final conclusion. These biases also have a failure rate and entrepreneurs make mistakes that lead to wrong decisions.

Busenitz and Barney found that the difference between entrepreneur's decision making and managers of larger organizations is the heuristics approach that most entrepreneurs follow.

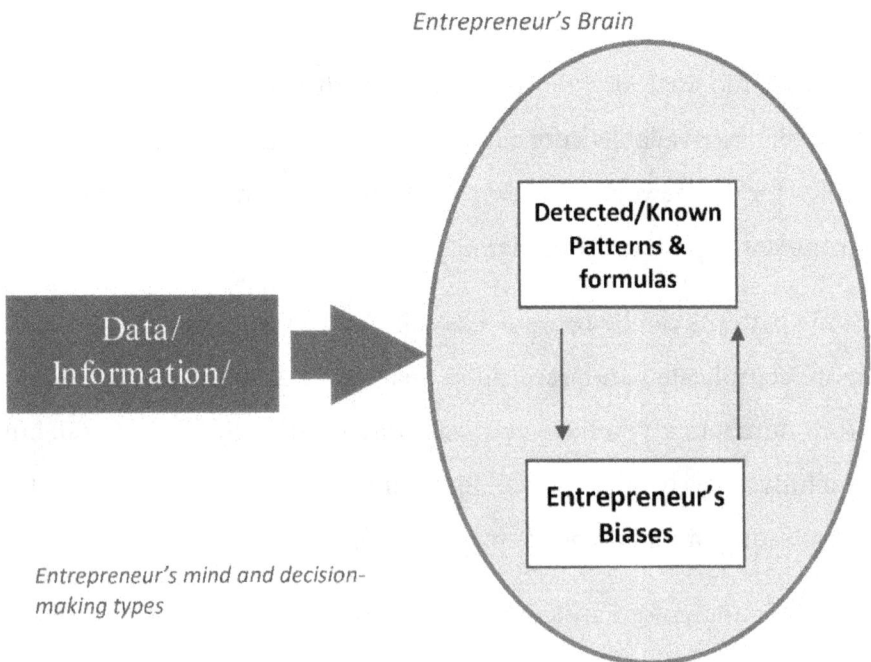

Entrepreneur's mind and decision-making types

On the other hand, managers are more keen to a more algorithmic decision making. There has been a myth about the fact that subjective approach is not going to lead to good results, while objective approach is considered as a wise and unbiased approach that has the best results.

The final decision of entrepreneurs, regardless of how many algorithms and patterns that are detected, is a subjective decision.

Elon Musk's bias towards going to space made some decisions that even CEOs of high-tech companies wouldn't make. Jeff Bezus also indicated that he made most of his decisions with his heart, guts and intuition rather than analysis.

As Baron and peer researchers found in their studies, connecting the dots is also crucial. Heuristics and algorithmic approaches of analysis require to be connected to each other and a feedback loop also exists in between.

Prior knowledge and background, trial and errors, biases, available information, and cognitive capability of entrepreneurs are all together in the analysis framework.

Entrepreneurial decisions are mostly finalized based on intuition rather than formulas, models and patterns, however, knowing patterns and models from data analysis would help entrepreneurs better plan for future circumstances.

The Golden Eagle Intelligence: The Brain of a Golden Eagle Entrepreneur

As mentioned, entrepreneurship is based on asymmetry in learning, knowledge, experience and cognitive capability of people. People with higher *Entrepreneurial Intelligence* are those individuals who possess a high degree of positive entrepreneurial asymmetry.

However, not every entrepreneur is a golden eagle entrepreneur. As we discovered in the previous chapters of this book, the golden eagle entrepreneur intelligence consists of a sum of eight unique traits that differentiates the golden eagle entrepreneur from the rest.

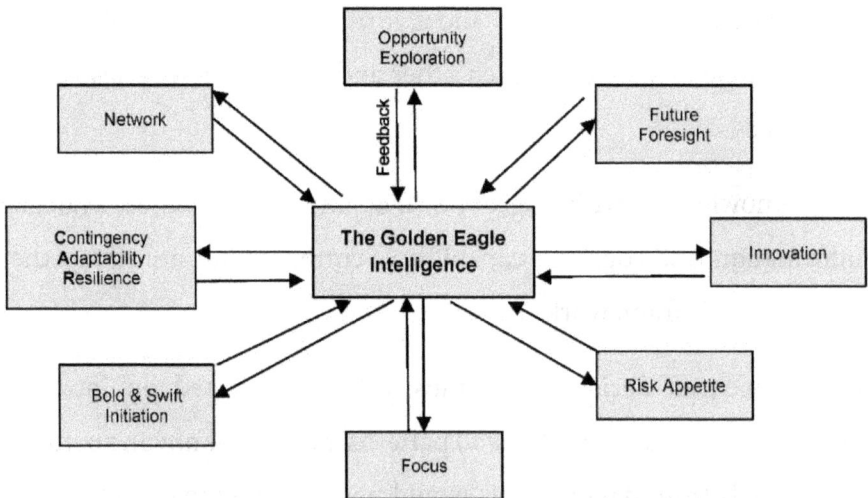

The golden eagle entrepreneur possesses the golden eagle intelligence and all the factors we covered in the previous chapters that work collectively together:

They are curious and while actively searching to *explore new opportunities*, they are also alert to new opportunities. They see opportunities that others won't recognize easily.

They possess a sharp eye that leads to *future foresight* and are ambitious enough to build the future instead of waiting for the future to happen.

They are not satisfied with the way things are done and embrace *innovation* to be change makers themselves. Their *risk appetite* makes them *focus* with determination and decisively take *bold and swift initiations*.

They build a vast *network* and leverage that network for future success. They accept *contingency*, uncertainty, and ambiguity and are *adaptable* enough to change their course of actions if it's needed. They are *resilient* and never let setbacks and obstacles stop them.

Finally, they use a balanced feedback loop to develop and refine what needs to be changed. Each element of the golden eagle intelligence - exploring opportunities, future foresight, innovation, risk appetite, focus, bold initiations, network, contingency, adaptability, resilience, failure, and feedback loop- is connected to golden eagle intelligence separately and collectively together they build the golden eagle intelligence inside the brain of the golden eagle entrepreneur. all actions and decisions of the golden eagle entrepreneur are controlled from this area.

Developing the Golden Eagle Intelligence

Soaring to the new heights of entrepreneurship requires a growth mindset. It's all about your brain and how you take care of your brain. Intuition and gut feelings are the main decision making methods inside the brain of the golden eagle entrepreneur.

To develop this type of brain, you need to focus on the building blocks of the golden eagle entrepreneur's brain and develop your brain accordingly.

- What are the current problems in our area of expertise or anywhere around us?
- What are the emerging trends in those areas?
- What opportunities are untouched in solving those problems?
- How is our industry in 10 years from now? What imaginary products and solutions could be there?
- What imaginary technologies can be developed to solve the current problems and disrupt current solutions?
- What are the unknown unknowns and how can we find them?
- What is the next innovation disrupting our industry?
- How can we make ourselves aware of the future risks?
- How can we mitigate the risks and embrace them?
- Are we aware of all risks around our business?
- What are the main objectives?
- What are distractions and noises and how can we separate them from the core of our business?
- Are short-term and long-term objectives aligned with each other?

- What bold action can we take?
- How can we improve the decision making process?
- Do we analyze too much before making decisions?
- How much do we trust our intuition and gut feelings in making decisions?
- How comfortable are we with imperfections?
- How comfortable we are in working in uncertain and ambiguous conditions?
- Are we prepared and aware of the setbacks and challenges that will come along the way?
- How determined and strong we are to stand against setbacks and challenges?
- Are we flexible enough to manage sudden changes in market, trends, or events?
- How do we build a network of key partners around us?
- How comfortable are we in embracing failure?
- Are we open to internal and external feedback?
- Where is the balance in the feedback loop and our intuition and guts?

By answering these questions or any tailored question from the building blocks of the golden eagle entrepreneur's brain, you may come up with yourself from reading this book, you can find out the golden eagle you are as an entrepreneur and score your own golden eagle intelligence.

FINAL THOUGHTS

"The Golden Eagle Entrepreneur" was penned with passion and dedication of time and we hope that our move towards bringing part of nature to the world of entrepreneurship can be beneficial for the future success of entrepreneurs.

Throughout the journey of this book, we tried to make an impact on the entrepreneurship ecosystem of the whole world by bringing a new approach.

Through our journey, we tried to demonstrate how the king of skies share common traits with some unique entrepreneurs.

We also maintain that this book is for founders, corporate leaders, venture capitalists, decision makers in the governments and public sector, as well as all educational institutions and even parents.

The concept can be used at school to train the mindset of students in entrepreneurship and give them a more golden eagle mindset. Founders can have a better picture of what it takes to be a golden eagle entrepreneur and VCs can use it as a framework to assess founders. Leaders can use the golden eagle strategy in their organization to help better sustainable future and long-term decisions while taking the main factors of the golden eagle intelligence in the developing strategies are also essential.

We believe that entrepreneurship is the answer to all problems in the business and that's why we believe that companies of any size can integrate a golden eagle intelligence into their organization.

All in all, we hope that you enjoyed this book and our effort would have been beneficial for building a better future and impact the whole world through innovation and entrepreneurial practices..

AUTHORS' BIO

Amir Salehi

Meet Amir Salehi, a seasoned entrepreneur with over 15 years of experience in significantly contributing to the field of startups, venture capital, corporate innovation, and accelerators. He studied Business Development and Internationalization at Umeå University in Sweden, with a focus on entrepreneurship.

Amir's career is marked by his leadership in establishing and leading VCs and corporate accelerators, showcasing his expertise in these areas. His passion for wildlife and the outdoors shapes his business philosophy, often drawing inspiration from nature to drive innovation.

As the founder and CEO of Nixaar, Amir is revolutionizing online commerce with a video and live commerce app builder that brings an in-store experience to customers' screens. Nixaar empowers sellers to connect with customers in a more human way, transforming the online shopping landscape.

Yasmin Ziaeian

Meet Yasmin Ziaeian, an accomplished academic with a PhD in Management. She lectures entrepreneurship and strategic management at Wroclaw University of Science and Technology in Poland, inspiring the next generation of business leaders.

With a wealth of experience in digital marketing, e-commerce, and agile methodologies, Dr. Ziaeian is a force to be reckoned with. She's a published researcher, renowned for her groundbreaking work on marketing strategies and technological innovation.

The following sources were used in writing this book

1. Abernikhina, I., 2023. Entrepreneurial Risk: Essence and Classification Depending on the Losses Borne by the Entrepreneur. *Oblik i finansi*, 100, pp.122-130.

2. Ackoff, R., 1989. From data to wisdom. *Journal of Applied Systems Analysis*, pp. 3-9.

3. Álvarez, A.M., García Merino, M.T. and Santos Álvarez, M.V., 2015. Information: The source of entrepreneurial activity. *Social Science Information*, 54(3), pp.280-298.

4. Alvarez, S.A. and Barney, J.B., 2007. Discovery and creation: Alternative theories of entrepreneurial action. *Strategic Entrepreneurship Journal*, 1(1-2), pp.11-26.

5. Ardichvili, A., Cardozo, R. and Ray, S., 2003. A theory of entrepreneurial opportunity identification and development. *Journal of Business Venturing*, 18(1), pp.105–123.

6. Ballantyne, D., Frow, P., Varey, R.J. and Payne, A., 2011. Value propositions as communication practice: Taking a wider view. *Industrial Marketing Management*, 40(2), pp.202-210.

7. Bambocci, A., 2003. Good to Great: Why Some Companies Make the Leap... And Others Don't. *Journal of Small Business Strategy (archive only)*, 14(2), pp.129-140.

8. Baron, R. A., 2006. Opportunity Recognition as Pattern Recognition: How Entrepreneurs "Connect the Dots" to Identify New Business Opportunities. *Academy of Management Perspectives*, 20(1), pp. 104-119.

9. Barrick, M.R. and Spilker, B.C., 2003. The relations between knowledge, search strategy, and performance in unaided and aided information search. *Organizational Behavior and Human Decision Processes*, 90(1), pp.1-18.

10. Baručić, A. and Umihanić, B., 2016. Entrepreneurship education as a factor of entrepreneurial opportunity recognition for starting a new business. *Management: Journal of Contemporary Management Issues*, 21(2), pp.27-44.

11. Basov, N. and Minina, V., 2018. Personal communication ties and organizational collaborations in networks of science, education, and business. *Journal of Business and Technical Communication*, 32(3), pp.373-405.

12. Battistella, C., De Toni, A.F. and Pessot, E., 2017. Open accelerators for start-ups success: a case study. *European Journal of Innovation Management*, 20(1), pp.80-111.

13. Bigelow, D., 2012. *The Start-Up of You: Adapt to the Future, Invest in Yourself, and Transform Your Career.* New York: Crown Publishing.

14. Blank, S. and Dorf, B., 2020. *The startup owner's manual: The step-by-step guide for building a great company.* John Wiley & Sons.

15. Blattenberger, G., Hyde, W.F. and Mills, T.J., 1984. Risk in fire management decision making: techniques and criteria. *Gen. Tech. Rep. PSW-80.* Berkeley, Calif.: US Department of Agriculture, Forest Service, Pacific Southwest Forest and Range Exp. Stn.

16. Blessin, B., 1999. *Innovations-und Umweltmanagement in kleinen und mittleren Unternehmen: Eine theoretische und empirische Analyse.* Frankfurt: Peter Lang International Academic Publishers.

17. Burgelman, R. A., 1983. Corporate Entrepreneurship and Strategic Management: Insights from a Process Study. *Management Science*, 29(12), pp. 1349-1364.

18. Burgelman, R.A., 1983. A Process Model of Internal Corporate Venturing in the Diversified Major Firm. *Administrative Science Quarterly*, pp. 223-244.

19. Busenitz, L.W. and Barney, J.B., *Entrepreneurs and Managers in Large Organizations: Biases and Heuristics in Strategic Decision-Making.* Unpublished paper, University of Houston and Ohio State University.

20. Catmull, E. and Wallace, A., 2014. *Creativity, Inc.: Overcoming the Unseen Forces That Stand in the Way of True Inspiration.* New York: Random House.

21. Choi, Y.R., Lévesque, M. and Shepherd, D.A., 2008. When should entrepreneurs expedite or delay opportunity exploitation?. *Journal of Business Venturing*, 23(3), pp.333-355.

22. Christensen, C.M., 2013. *The Innovator's Dilemma: When New Technologies Cause Great Firms to Fail.* Harvard Business Review Press.

23. Coburn, D., 2014. *Networking is Not Working: Stop Collecting Business Cards and Start Making Meaningful Connections.* IdeaPress Publishing.

24. Collins, J., 2009. *Good to Great: Why Some Companies Make the Leap and Others Don't.* HarperCollins.

25. Cooper, A.C., Folta, T.B. and Woo, C.Y., 1995. Entrepreneurial Information Search. *Journal of Business Venturing,* 10(2), pp.107-120.

26. Corbett, A.C., 2007. Learning asymmetries and the discovery of entrepreneurial opportunities. *Journal of Business Venturing,* 22(1), pp.97–118.

27. Corbo, L., Kraus, S., Vlačić, B., Dabić, M., Caputo, A. and Pellegrini, M.M., 2023. Coopetition and innovation: A review and research agenda. *Technovation,* 122, p.102624.

28. Croll, A. and Yoskovitz, B., 2013. *Lean Analytics: Use Data to Build a Better Startup Faster.* Sebastopol: O'Reilly Media, Inc.

29. Daniel, K., 2017. *Thinking, Fast and Slow.* New York: Farrar, Straus and Giroux.

30. Davis, J., Wolff, H.G., Forret, M.L. and Sullivan, S.E., 2020. Networking via LinkedIn: An examination of usage and career benefits. *Journal of Vocational Behavior,* 118, p.103396.

31. Dickinson, G., 2013. *Eagle, Birds of Prey: A Book with Facts and Pictures.* N/A: CreateSpace Independent Publishing Platform.

32. Doerr, J., 2018. *Measure What Matters: How Google, Bono, and the Gates Foundation Rock the World with OKRs.* Penguin.

33. Dunne, P. and Karlson, K.T., 2017. *Birds of Prey: Hawks, Eagles, Falcons, and Vultures of North America.* Houghton Mifflin Harcourt.

34. Dweck, C.S., 2006. *Mindset: The New Psychology of Success.* New York: Random House.

35. Dyer, J., Gregersen, H. and Christensen, C.M., 2019. *The Innovator's DNA, Updated, with a New Preface: Mastering the Five Skills of Disruptive Innovators.* Harvard Business Press.

36. Earnest, P. and Karinch, M., 2010. *Business Confidential.* New York: AMACOM.

37. Eckhardt, J.T. and Shane, S.A., 2003. Opportunities and Entrepreneurship. *Journal of Management*, 29(3), pp.333–349.

38. El hombre y la Tierra. 1974. [Film] Directed by Félix Rodríguez de la Fuente. Spain: s.n.

39. Ellis, S. and Brown, M., 2017. *Hacking Growth: How Today's Fastest-Growing Companies Drive Breakout Success.* Currency.

40. Eyal, N., 2014. *Hooked: How to Build Habit-Forming Products.* Penguin.

41. Fairlie, R.W. and Fossen, F.M., 2018. Opportunity versus necessity entrepreneurship: Two components of business creation. *Small Business Economics*, 51, pp.609-629.

42. Fang, X. and An, L., 2017. A study of effects of entrepreneurial passion and risk appetite on entrepreneurial performance. *Revista de Cercetare si Interventie Sociala*, 56, pp.102-113.

43. Fernández, C., Guerrero, M. and Urbano, D., 2015. Business incubation: innovative services in an entrepreneurship ecosystem. *The Service Industries Journal*, 35(7-8), pp.783–800.

44. Ferriss, T., 2011. *The 4-Hour Work Week: Escape the 9-5, Live Anywhere and Join the New Rich.* New York: Random House.

45. Foss, N.J., Lyngsie, J. and Zahra, S.A., 2013. The role of external knowledge sources and organizational design in the process of opportunity exploitation. *Strategic Management Journal*, 34(12), pp.1453-1471.

46. Frese, M. and Gielnik, M.M., 2023. The psychology of entrepreneurship: action and process. *Annual Review of Organizational Psychology and Organizational Behavior*, 10(1), pp.137-164.

47. Fried, J. and Hansson, D.H., 2010. *Rework.* New York: Crown Currency.

48. Gaglio, C.M., 2004. The Role of Mental Simulations and Counterfactual Thinking in the Opportunity Identification Process. *Entrepreneurship Theory and Practice*, 28(6), pp.1042-2587.

49. Gerber, M.E., 2021. *The E-Myth Revisited: Why Most Small Businesses Don't Work and What to Do About It.* HarperCollins Publishers Ltd.

50. Gerig, V., 1998. *Kriterien zur Beurteilung unternehmerischen Handelns von Mitarbeitern und Führungskräften.* München und Mering: Rainer Hampp Verlag.

51. Godin, S., 2008. *Tribes: We Need You to Lead Us.* Penguin.

52. Gordon, G. and Nelke, A., 2017. *CSR und nachhaltige Innovation: Zukunftsfähigkeit durch soziale, ökonomische und ökologische Innovationen.* Berlin: Springer.

53. Grant, A., 2014. *Give and Take: Why Helping Others Drives Our Success.* Penguin.

54. Guillebeau, C., 2012. *The $100 Startup: Reinvent the Way You Make a Living, Do What You Love, and Create a New Future.* Crown Currency.

55. Heath, C. and Heath, D., 2007. *Made to Stick: Why Some Ideas Survive and Others Die.* Random House.

56. Hilbert, M., 2016. Formal definitions of information and knowledge and their role in growth through structural change. *Structural Change and Economic Dynamics,* 37, pp.69–82.

57. Hitt, M.A., Ireland, R.D., Camp, S.M. and Sexton, D.L., 2001. Strategic Entrepreneurship: Entrepreneurial Strategy for Wealth Creation. *Strategic Management Journal,* 22, pp.479–491.

58. Hitt, M.A., Ireland, R.D., Sirmon, D.G. and Trahms, C.A., 2011. Strategic Entrepreneurship: Creating Value for Individuals, Organizations, and Society. *Academy of Management Perspectives,* 25(2), pp.57-75.

59. Hmieleski, K.M. and Baron, R.A., 2008. Regulatory focus and new venture performance: A study of entrepreneurial opportunity exploitation under conditions of risk versus uncertainty. *Strategic Entrepreneurship Journal,* 2(4), pp.285-299.

60. Ireland, R.D., Hitt, M.A. and Sirmon, D.G., 2003. A Model of Strategic Entrepreneurship: The Construct and its Dimensions. *Journal of Management,* 29(6), pp.963–989.

61. Ireland, R.D. and Webb, J.W., 2007. Strategic Entrepreneurship: Creating Competitive Advantage through Streams of Innovation. *Business Horizons,* 50(1), pp.49-59.

62.   Jemal, S., 2020. Effect of entrepreneurial mindset and entrepreneurial competence on performance of small and medium enterprise, evidence from literature review. *International Journal of Management & Entrepreneurship Research*, 2(7), pp.476-491.

63.   John, D. and Paisner, D., 2016. *The Power of Broke: How Empty Pockets, a Tight Budget, and a Hunger for Success Can Become Your Greatest Competitive Advantage*. Crown Currency.

64.   Jothi, P.S., Neelamalar, M. and Prasad, R.S., 2011. Analysis of social networking sites: A study on effective communication strategy in developing brand communication. *Journal of Media and Communication Studies*, 3(7), pp.234-242.

65.   Kahneman, D. and Lovallo, D., 1993. Timid Choices and Bold Forecasts: A Cognitive Perspective on Risk Taking. *Management Science*, 39(1), pp.17-31.

66.   Kahneman, D., 2011. *Thinking, Fast and Slow*. New York: Farrar, Straus and Giroux.

67.   Kerley, L.L. and Slaght, J.C., 2013. First documented predation of Sika deer (Cervus nippon) by Golden Eagle (Aquila chrysaetos) in the Russian far east. *Journal of Raptor Research*, 47(3), pp.328-330.

68.   Khan, S., 2013. Mapping entrepreneurship ecosystem of Saudi Arabia. *World Journal of Entrepreneurship, Management and Sustainable Development*, 9(1), pp.28-54.

69.   Kleinschmidt, E.J., Geschka, H. and Cooper, R.G., 2013. *Erfolgsfaktor Markt: Kundenorientierte Produktinnovation*. Berlin: Springer-Verlag.

70.   Knapp, J., Zeratsky, J. and Kowitz, B., 2016. *Sprint: How to Solve Big Problems and Test New Ideas in Just Five Days*. New York: Simon and Schuster.

71.   Koehn, N., 2014. *The Hard Thing About Hard Things: Building a Business When There Are No Easy Answers*. Harper Business.

72.   Kohnen, J., 2008. *Crucial Conversations: Tools for Talking When Stakes Are High*. McGraw-Hill Education.

73.   Krackhardt, D., Nohria, N. and Eccles, R.G., 2003. The strength of strong ties. In: R. Cross, A. Parker and L. Sasson, eds., *Networks in the Knowledge Economy*, 1st ed. New York: Oxford University Press, pp.82-108.

74. Kumar, A., Manjunath, D. and Kuri, J., 2004. *Communication Networking: An Analytical Approach*. Morgan Kaufmann.

75. Kuratko, D.F., Montagno, R.V. and Hornsby, J.S., 1990. Developing an Intrapreneurial Assessment Instrument for an Effective Corporate Entrepreneurial Environment. *Strategic Management Journal*, 11(5), pp.49-58.

76. Kuratko, D.F., Morris, M.H. and Covin, J.G., 2015. Corporate entrepreneurship: the innovative challenge for a new global economic reality. *Springer Science+Business Media*, 21(3), pp.245–253.

77. Lange, G.S. and Johnston, W.J., 2020. The value of business accelerators and incubators–an entrepreneur's perspective. *Journal of Business & Industrial Marketing*, 35(10), pp.1563-1572.

78. Lawson, M., 2014. The impact of risk propensity on corporate entrepreneurship. *Entrepreneurship Research Journal*, 4(4), pp.315-332.

79. Lesch, M.E., Dach, M. and Brekke, D., 2005. *Never Eat Alone: And Other Secrets to Success, One Relationship at a Time*. New York: Crown Business.

80. López-Muñoz, J.F., Novejarque-Civera, J. and Pisá-Bó, M., 2023. Innovative entrepreneurial behavior in high-income European countries. *International Journal of Entrepreneurial Behavior & Research*, 29(7), pp.1516-1540.

81. Love, D. and Watson, M., 1990. *The Golden Eagle*. Buckinghamshire: Thomas & Sons.

82. Lusch, R.F. and Nambisan, S., 2015. Service innovation. *MIS Quarterly*, 39(1), pp.155-176.

83. Lussier, R.N., Sonfield, M.C. and Lussier, R.A., 2001. Strategies Used by Small Business Entrepreneurs. *American Journal of Business*, 16(1), pp.29-38.

84. Mackay, H., 1999. *Dig Your Well Before You're Thirsty: The Only Networking Book You'll Ever Need*. Crown Currency.

85. March, J.G., 1991. Exploration and exploitation in organizational learning. *Organization Science*, 2(1), pp.71-87.

86. McNally, D., 1999. *The Eagle's Secret: Success Strategies for Thriving at Work and in Life*. Dell.

87. Meyer, R., 2008. *Die Entwicklung des Betriebswirtschaftlichen Risiko-und Chancenmanagements.* Berlin: Springer-Verlag.

88. Miller, D.J., Fern, M.J. and Cardinal, L.B., 2007. The use of knowledge for technological innovation within diversified firms. *Academy of Management Journal,* 50(2), pp.307-325.

89. Mintzberg, H., 1978. Patterns in Strategy Formation. *Management Science,* 24(9), pp.934-948.

90. Misner, I. and Hilliard, B., 2017. *Networking like a Pro: Turning Contacts into Connections.* Entrepreneur Press.

91. Moore, G.A. and McKenna, R., 1999. *Crossing the Chasm: Marketing and Selling Disruptive Products to Mainstream Customers.* Harper Business.

92. Murray, A.I., 1984. A Concept of Entrepreneurial Strategy. *Strategic Management Journal,* 5(1), pp.1-13.

93. Naumann, C., 2017. Entrepreneurial mindset: A synthetic literature review. *Entrepreneurial Business and Economics Review,* 5(3), pp.149-172.

94. Ncokazi, A. and Mpiti, P.T., 2023. Critical evaluation of the relationship between the need for achievement and entrepreneurship performance: risk-taking propensity, entrepreneurial independence and motivation. In: C. Mihaela and D. Silvia, eds., *Motivation and Success,* 1st ed. London: IntechOpen.

95. Oakey, R., 2003. Technical entrepreneurship in high technology small firms: some observations on the implications for management. *Technovation,* 23(8), pp.679–688.

96. Olsen, D., 2015. *The Lean Product Playbook: How to Innovate with Minimum Viable Products and Rapid Customer Feedback.* John Wiley & Sons.

97. Olsen, L.E., 2008. Professional networking online. *A qualitative study of LinkedIn use in Norway* (Master's thesis, The University of Bergen).

98. Ortega Álvarez, A.M., García Merino, M.T. and Santos Álvarez, M.V., 2015. Information: The source of entrepreneurial activity. *Social Science Information,* 54(3), pp.280-298.

99. Osterwalder, A. and Pigneur, Y., 2010. *Business Model Generation: A Handbook for Visionaries, Game Changers, and Challengers.* John Wiley & Sons.

100. Parker, S.C., 2001. Intrapreneurship or entrepreneurship?. *Journal of Business Venturing*, 16(1), pp.19–34.

101. Piscione, D.P., 2014. *Secrets of Silicon Valley: What Everyone Else Can Learn from the Innovation Capital of the World.* New York: Palgrave Macmillan.

102. Piscione, D.P., 2014. *Secrets of Silicon Valley: What Everyone Else Can Learn from the Innovation Capital of the World.* New York: St. Martin's Press. Kindle Edition.

103. Pink, D.H., 2011. *Drive: The Surprising Truth About What Motivates Us.* Penguin.

104. Preston, C.R., 2004. *Golden Eagle: Sovereign of the Skies.* Oregon: Graphic Arts Center Pub.

105. Preston, C.R. and Leppart, W.J., 2004. *The Golden Eagle: Sovereign of the Skies.* Oregon: Graphic Arts Center Publishing Company.

106. Ramoglou, S. and McMullen, J.S., 2024. "What is an opportunity?": From theoretical mystification to everyday understanding. *Academy of Management Review*, 49(2), pp.273-298.

107. Ratten, V., 2023. Entrepreneurship: Definitions, opportunities, challenges, and future directions. *Global Business and Organizational Excellence*, 42(5), pp.79-90.

108. Rejeb-Khachlouf, N., Mezghani, L. and Quélin, B., 2011. Personal networks and knowledge transfer in inter-organizational networks. *Journal of Small Business and Enterprise Development*, 18(2), pp.278-297.

109. Rice, M.P., Fetters, M.L. and Greene, P.G., 2014. University-based entrepreneurship ecosystems: a global study of six educational institutions. *International Journal of Entrepreneurship and Innovation Management*, 18(5/6), pp.481-501.

110. Ries, E., 2011. *The Lean Startup: How Today's Entrepreneurs Use Continuous Innovation to Create Radically Successful Businesses.* New York: Crown Publishing.

111. Robbe, M. and Nathan, S., 2014. Appreciation: Proposal of a new concept in business discipline. *Business Horizons*, 57(3), pp.343-353.

112. Rowley, J., 2000. From learning organization to knowledge entrepreneur. *Journal of Knowledge Management*, 4(1), pp.7-15.

113. Rowley, J., 2007. The wisdom hierarchy: representations of the DIKW hierarchy. *Journal of Information Science*, 33(2), pp.163–180.

114. Russell, R.A. and Russell, J.R., 1992. An Examination of the Effects of Organizational Norms, Organizational Structure, and Environmental Uncertainty on Entrepreneurial Strategy. *Journal of Management*, 18(4), pp.639-656.

115. Salancik, G.R., Krackhardt, D., Andrews, S.B. and Tolbert, P.S., 1995. From the book review editor--Structural Holes: The Social Structure of Competition by Ronald S. Burt. *Administrative Science Quarterly*, 40(2), p.343.

116. Sandberg, S., 2015. *Lean In: Women, Work, and the Will to Lead*. New York: Knopf.

117. Sarasvathy, S.D., 2001. Causation and Effectuation: Toward a Theoretical Shift from Economic Inevitability to Entrepreneurial Contingency. *Academy of Management Review*, 26(2), pp.243-263.

118. Sauers, A., 2008. *Effective Customer Relationship Management*. Cambria Press.

119. Savage, C.S., 1987. *Eagles of North America*. New York: Crown Publishers.

120. Selig, G., 2014, April. Critical success factors for winning entrepreneurs and the role of an incubator in accelerating the growth of start-ups and early stage companies. In: *Proceedings of the 2014 Zone 1 Conference of the American Society for Engineering Education*, pp.1-6. IEEE.

121. Shamsudeen, K., Keat, O.Y. and Hassan, H., 2017. Entrepreneurial success within the process of opportunity recognition and exploitation: An expansion of entrepreneurial opportunity recognition model. *International Review of Management and Marketing*, 7(1), pp.107-111.

122. Shane, S. and Venkataraman, S., 2000. The Promise of Entrepreneurship as a Field of Research. *The Academy of Management Review*, 25(1), pp.217-226.

123. Shane, S., 2000. Prior Knowledge and the Discovery of Entrepreneurial Opportunities. *Organization Science*, 11(4), pp.448-469.

124. Shannon, C.E., 1948. A Mathematical Theory of Communication. *Bell Systems Technical Journal*, 27(3), pp.379-423.

125. Silva, D.S., Ghezzi, A., Aguiar, R.B.D., Cortimiglia, M.N. and ten Caten, C.S., 2021. Lean startup for opportunity exploitation: adoption constraints and strategies in technology new ventures. *International Journal of Entrepreneurial Behavior & Research*, 27(4), pp.944-969.

126. Sinek, S., 2009. *Start with Why: How Great Leaders Inspire Everyone to Take Action*. Penguin.

127. Singh, B., 2016. *The Lean Startup: How Today's Entrepreneurs Use Continuous Innovation to Create Radically Successful Businesses*. John Wiley & Sons.

128. Sonfield, M.C. and Lussier, R.N., 2000. Innovation, risk and entrepreneurial strategy. *Entrepreneurship and Innovation*, 1(2), pp.91-97.

129. Steenkamp, A., Meyer, N. and Bevan-Dye, A.L., 2024. Self-esteem, need for achievement, risk-taking propensity and consequent entrepreneurial intentions. *The Southern African Journal of Entrepreneurship and Small Business Management*, 16(1), p.753.

130. Stephan, U., 2018. Entrepreneurs' mental health and well-being: A review and research agenda. *Academy of Management Perspectives*, 32(3), pp.290-322.

131. Stewart Jr, W.H. and Roth, P.L., 2001. Risk propensity differences between entrepreneurs and managers: A meta-analytic review. *Journal of Applied Psychology*, 86(1), pp.145-153.

132. Stowell, S.J. and Mead, S.S., 2016. *The Art of Strategic Leadership: How Leaders at All Levels Prepare Themselves, Their Teams, and Organizations for the Future*. John Wiley & Sons.

133. Taleb, N.N., 2007. *The Black Swan: The Impact of the Highly Improbable*. New York: Random House.

134. Tang, J., Kacmar, K.M. and Busenitz, L.W., 2012. Entrepreneurial alertness in the pursuit of new opportunities. *Journal of Business Venturing*, 27(1), pp.77–94.

135. Teberga, P.M.F. and Oliva, F.L., 2018. Identification, analysis and treatment of risks in the introduction of new technologies by start-ups. *Benchmarking: An International Journal*, 25(5), pp.1363-1381.

136. Thiel, P. and Masters, B., 2014. *Zero to One: Notes on Startups, or How to Build the Future.* New York: Crown Currency.

137. Thornberry, N., 2001. Corporate Entrepreneurship: Antidote or Oxymoron?. *European Management Journal*, 19(5), pp.526–533.

138. Tyrkalo, Y., 2022. Entrepreneurial Risks: Causes, Consequences and Management (Theoretical Aspects). *Path of Science*, 8(1), pp.3010-3017.

139. Ucbasaran, D., Westhead, P. and Wright, M., 2009. The extent and nature of opportunity identification by experienced entrepreneurs. *Journal of Business Venturing*, 24(2), pp.99–115.

140. Uzunidis, D. and Saulais, P. eds., 2017. *Innovation Engines: Entrepreneurs and Enterprises in a Turbulent World.* John Wiley & Sons.

141. Uzzi, B., 2018. Social structure and competition in interfirm networks: The paradox of embeddedness. In: M. Granovetter, ed., *The Sociology of Economic Life*, 1st ed. London: Routledge, pp.213-241.

142. Vance, A., 2015. *Elon Musk: Tesla, SpaceX, and the Quest for a Fantastic Future.* HarperCollins. Kindle Edition.

143. Vaghely, I.P. and Julien, P.A., 2010. Are opportunities recognized or constructed? An information perspective on entrepreneurial opportunity identification. *Journal of Business Venturing*, 25(1), pp.73–86.

144. Vesković, N., 2014. Aspects of entrepreneurial risk. *FINIZ 2014-The Role of Financial Reporting in Corporate Governance*, pp.115-117.

145. Vonortas, N.S. and Kim, Y., 2015. Managing risk in new entrepreneurial ventures. In: *Dynamics of Knowledge Intensive Entrepreneurship*, 1st ed. London: Routledge, pp.145-165.

146. Walker, M., 2017. *Why We Sleep: Unlocking the Power of Sleep and Dreams.* New York: Scribner.

147. Walling, R., 2010. *Start Small, Stay Small: A Developer's Guide to Launching a Startup.* The Numa Group LLC.

148. Watson, J., 2010. *The Golden Eagle*. London: Bloomsbury Publishing.

149. Willink, J. and Babin, L., 2015. *Extreme Ownership: How U.S. Navy SEALs Lead and Win*. New York: St. Martin's Press.

150. Yampolskiy, R.V., 2016. Efficiency Theory: A Unifying Theory for Information, Computation and Intelligence. *Journal of Discrete Mathematical Sciences and Cryptography*, 19(2), pp.259-277.

151. Yi-xin, Z., 2006. An investigation on the theory of information-knowledge-intelligence transforms. *Front. Electr. Electron. Eng*, 1(4), pp.400–404.

152. Zahra, S.A. and Nambisan, S., 2011. Entrepreneurship in global innovation ecosystems. *Academy of Management Perspectives*, 25(4), pp.4-17.

153. Zemlyak, S., Naumenkov, A. and Khromenkova, G., 2022. Measuring the Entrepreneurial Mindset: The Motivations behind the Behavioral Intentions of Starting a Sustainable Business. *Sustainability*, 14(23), p.15997.

154. Zimmerman, J. and Ng, D., 2015. *Social Media Marketing All-in-One for Dummies*. John Wiley & Sons.

155. https://www.nationalgeographic.com/animals/birds/g/golden-eagle

www.ingramcontent.com/pod-product-compliance
Lightning Source LLC
Chambersburg PA
CBHW071206210326
41597CB00016B/1690